ONE
COSMIC
INSTANT

ONE COSMIC INSTANT

A Natural History
of Human Arrogance

JOHN A. LIVINGSTON

McClelland and Stewart Limited

The Canadian Publishers
McClelland and Stewart Limited
25 Hollinger Road, Toronto

Printed and bound in Canada

Portions of Chapter Two, "One in a Million," in different form appeared in *Canada*, by John A. Livingston (Natural Science of Canada Limited, 1970).

For Peggy
Sally, Peter, John

ACKNOWLEDGMENTS

Many of the notions in this book had been nagging at me for expression for an unconscionably long time. A sufficient period in which to sort and begin to shape them would probably never have been forthcoming without a special grant generously contributed by the Canadian National Sportsmen's Show. My gratitude to the C.N.S.S. and to its president, Mr. R. T. D. Birchall, is warmly acknowledged.

Further essential support for the completion of the book was made available to me through the initiative of Dr. C. David Fowle, who with his colleagues at York University, Dr. G. A. P. Carrothers and Dr. Robert Haynes, devised interesting and most rewarding ways and means of keeping the project alive.

Judith Symons' library work was indispensable, as was thoughtful, sympathetic and patient editing by Lily Miller.

John Livingston
Faculty of Environmental Studies
York University
1972

CONTENTS

AUTHOR'S FOREWORD

Frustration, dismay, and varying degrees of anger have been the inseparable companions of the wildlife conservationist always. It is a fine line between these emotions and misanthropy. Rather than opt for troglodytic retirement, however, I have chosen the route toward what I consider a reasonably civilized form of sabotage.

In the course of writing this, I have discovered that buildings and mechanical contrivances must be a great deal more susceptible to explosion than are concreted ideas. Most particularly resistant is the edifice which is collective Western homocentricity, bulwarked and reinforced as it is by apathy, greed, ignorance – and tradition.

Although I am very much aware that there is nothing one can know for certain, I have proceeded from one radical assumption. Contemporary Western man in the overwhelming majority considers himself fundamentally different and distinct from the living world that gives him both substance and sustenance. This imagined separation between man and "nature" has provided the conceptual framework for a further doctrine, that of absolute human power and authority over the nonhuman. These ludicrous but terrifying notions have become solidified in our collective thought in a ridiculously brief period of human and Earth history.

The conceptual power structure is recalcitrant. Its successful penetration and demolition will be essential to the continuance of life processes as they presently exist. I have attempted to illuminate in perspective a few of the more obvious building-blocks between which the fuses must be insinuated.

CHAPTER ONE

The Temple of the Maya

Massive cream-and-sable king vultures have been wheeling above the Temple of the Giant Jaguar for centuries. Red, blue, and orange-carunculated heads cocked pale eyes downward at the great gleaming plaza of Tikal, where a brilliant civilization shone momentarily, only to be reclaimed by the oppressive lowland jungle of Guatemala.

The Maya knew the vulture, and carved his likeness on stone. Oc, the vulture, knew the Maya, and he watched the Maya vanish. The vulture still circles overhead. Parrots screech and toucans croak from the canopy of towering mahoganies, and ocellated turkeys gingerly make their way across openings on the forest floor. Except for the absence of the Maya and the emptiness of his shrine, little has changed.

In a cosmic sense, nothing has changed at Tikal, for civilization is ephemeral. Silently the king vulture drifts across a gap in the forest cover around the ruins. Once, a culture as sophisticated as the world had ever seen flourished here. During Europe's Dark Ages, the Maya knew about astronomy and mathematics, and had a concept of time.

But cultures are fleeting. They evolve – sometimes with extraordinary suddenness – and wither just as quickly. On the other hand, the king vulture was the same animal before, during, and after the brief scintillating moment of the Maya.

I – a birdwatcher – left Guatemalan Mayaland with far more than a list of new tropical species. I left with a sense, not so much of the vulnerability of human cultures, but of their susceptibility to unexpected, rapid change. In thinking of the relative immutability of the king vulture as a species over a time-span of mere centuries, I was struck by the vast difference between the slowness of natural selection and the speed of the rise and fall of cultures. I came to brood on human aspirations, and on their potential for taking new directions.

If the naturalist values the objects of his obsession, he is compelled to involve himself in the world of people who at this moment are watching, with indifference, as the king vulture and his tribe gradually disappear. Like that of the Maya, something in our society is grievously wrong.

The nonhuman world is important to the birdwatcher. Its relative importance grows, in inverse relationship with an inevitable misanthropy. One's disillusionment with society's treatment of nonhuman nature is built on a body of evidence which is conspicuous on every hand.

One of the commonest questions I am asked is: "What 'turned you on' to birds?" When I reply, truthfully, that I cannot remember, I am rarely believed. People look for instant, simple causes for observed effects. It was too long ago for me to be able to recall; certainly it was before I was ten years old. Perhaps it is more enlightening to say that I cannot remember when I was *not* interested in birds.

Children respond to animals, and as a youngster I realized that a ravine, then unspoiled, near my house was full of wild animals, especially during spring and fall bird migrations. At those times I would conceal myself under a shrub, remain as inconspicuous as possible, and wait for the birds to come to me. But I had great trouble attaching names to them; they moved too quickly for me to pick up their distinguishing features. In my school library I had already discovered the stunning array of Fuertes and Brooks plates in Forbush's *Birds of Massachusetts* ; I had discovered Frank Chapman, and Reed's little guides, but all of these had stern limitations for field identification. The living world of birds was explosively revealed to me in 1934, when I was eleven years old. Like Keats opening Chapman's Homer, I first looked that year into Roger Peterson's earliest *Field Guide to the Birds.* My response was not so much like that of someone watching a new planet swim into his ken as like one liberated into a new galaxy, equipped with the tools with which to comprehend it.

I especially remember one May morning, during a warbler "wave." That day there was an unusual number and variety of warblers. With Peterson's book in hand, I hid motionless in my retreat and watched the parade. I was surrounded by birds – darting, fidgeting, flashing, shimmering – a magnolia warbler, a black-throated blue, a black-throated green, a myrtle! And I could now *identify* the birds, which brought me infinitely closer to them.

From the ability to attach names to the hundreds of kinds of birds, plants, butterflies and other forms of life flows the realization that these new phenomena have about them a degree of predictability. One doesn't look for the brilliant larvae of the monarch butterfly outside the milkweed patch, nor does one tune an ear for the song of a robin until late in March. One looks in beech trees for the nesting excavations of pileated woodpeckers, and discovers that a tiny nest with an inner lining of horse-hair is probably that of a chipping sparrow.

It requires little ecological insight to learn that one must go to marshes to find bitterns, rails, swamp sparrows and marsh wrens, and to open fields and meadows for bobolinks and savannah sparrows. It becomes apparent that there are specific habitat requirements for birds. The moment some favorite location for a species is changed or eliminated by human activity, there are issues at stake. The boy who saw Cory's least bittern (a rare and local color phase) driven from the marshes of Ashbridge's Bay in Toronto by a sorely needed sewage treatment plant learned early of the complexities and conflicts of conservation.

All is not orchard orioles and apple blossoms, nor purple finches at the backyard feeder. Though we are aware of human over-population and the attendant problems, not until their effects are seen first-hand are we struck by the immediacy and reality of the modern environmental crisis. One rages inwardly at the effects of sheer human numbers on the landscapes and the birds one wishes to preserve, and laments the degradation of the human condition. Awareness of what is happening to the things one holds precious leads the observer to question the human "values" which have brought the world of nature to its present level of subjugation. The things I value – such as birds – are being destroyed by other

things I also value: human life. The conflicts and the para-
doxes are staggering.

Not all in the birdwatcher's life is grimness and dismay.
There is the lighter side. As his travels become more exten-
sive and his virtuosity develops, he becomes able to practice
ornithological "one-upmanship." While observing a gray-
headed gull in the humid estuary of Ecuador's Guyas River,
he might murmur, "Ah. Last time I saw that species was on
Lake George in Uganda, under the Mountains of the Moon."
Or, "In the forest of Bialowieza in eastern Poland you can see
all four species of European flycatchers in one day." Or,
"Isn't it odd – the discontinuity in the distribution of the
azure-winged magpie, which I have seen in both Portugal and
Japan. Except for Mongolia and northern China, it does not
occur anywhere in between."

More important than the birdwatcher's skill and knowledge is his awareness of the cyclic and repeating nature of life.
The seasons are recurring and predictable, but things are
never *exactly* the same from year to year. As I write this, it is
mid-March. Outside my window a supercharged cock house
sparrow has begun his courtship ritual, and is trembling with
intensity. There is a small flock of slate-colored juncos on the
dirty snow near the back fence, foraging beneath reddening
osier canes. A cardinal is singing lustily a couple of gardens
away, and the backyard skating rink is melting fast. Two or
three crows have gone by in the last few minutes, the advance
guard of regiments to come. There is a great flutter of copulating pigeons on a chimney-top across the way. The day is
overcast, but the prospect is good. Any day I will hear the
grinding scrape of a grackle's song, and my faith in the continuity of things will be reaffirmed.

This spring, predictable though it may be, will not be like
any I have seen before. No two springs are ever alike; things
change in nature, and flux does not end with the circling
seasons. One cannot anticipate everything. Still, I am sad-
dened by the certain knowledge that this spring there will be
fewer mallards and canvasbacks than there were in other

years. There will be fewer blue whales – almost none, in fact. There will not be as many leopards or pelicans as there were in the springs of my youth. I will probably not see a peregrine falcon this year. Because of real estate development on their wintering grounds in the Bahamas, there will be fewer Kirtland's warblers in Michigan than there were a year ago. But there will be seventy million more people on Earth than there were at this time last year.

One is drawn into conservation in any number of ways. A farmer becomes aware of soil erosion and soil depletion and water shortages; an angler becomes aware of water pollution and eutrophication, hydro dams, and over-fishing; a bird-watcher notices the steady disappearance of the habitats that sustain his objects of interest, and the poisoning of those natural habitats that do remain. The common factor in all of these environmental changes is man.

Something in the biosphere is drastically out of synchronization with everything else. That *something* is man. One need not be a prophet of doom, nor even a mildly conventional pessimist, to see that there are components in the biospheric life system which are dangerously out of hand, and that time to restore the bond with nature is limited.

I have spent a lifetime in conservation affairs and have gradually but inescapably been convinced that up until the present time all we conservationists have been able to achieve is fight a delaying action. The best we have been able to accomplish has been the temporary treatment of symptoms. We have not reached the root cause of environmental distress, which is Western industrial man.

Western civilized society appears to be indifferent toward nonhuman nature. We have seen the bison, the trumpeter swan, and the bighorn sheep fall before the gunners; we have seen the prairie dog, the black-footed ferret and the whooping crane give way before the sod-busters; we have seen the giant baleen whales reduced to the vanishing point by international commercial greed. Most significant of all, perhaps, has been the unchanging traditional assumption that although the loss of these animals may well have been regrettable, it was inevitable and unavoidable in the context of the advancement of human progress. It was sad that these things had to happen but, after all, what was at stake was human welfare, and in the final analysis human welfare comes first. That's only "natural."

The World's Fair at Montreal, "Expo '67," was subtitled *Man and His World.* To this day its amusement-park successor remains *Terre des Hommes.* In 1967, my thoughts on man and his world drifted back to Nicolaus Copernicus.

Nicolaus Copernicus was born about 500 years ago, and in the course of his life he demonstrated the theory that the sun – not man – is at the center of our planetary system, and that the Earth and the other planets revolve around it. His system was later developed by the likes of Kepler, Galileo, and Newton. However, Copernicus had merely resurrected and improved upon a notion that had occurred to Pythagoras of Samos, who was born more than 2,500 years ago. Yet, the theories of Copernicus and Pythagoras have had little impact on modern man, who still terms it, "Man and His World."

This reflects an ancient and profound view with regard to the role of the nonhuman component of the biosphere. It is the root cause of the despoliation and disfigurement of nonhuman nature which has now become the human environmental crisis. The conservation movement is doomed to failure unless Western man summons the courage and the will to challenge his fundamental assumptions.

Nowadays, public attention given to environmental questions generally concerns pollution – the contamination or overloading of air, water, soil, and living tissue by both natural and synthetic substances in volume or of kind with which natural systems cannot cope. This is a symptom of a much more deep-seated illness. Pollution *can* be dealt with. The *causes* of pollution are far more difficult to approach. Pollution and its effects constitute a rash of conspicuous abscesses which are the external manifestations of something mortally wrong within.

It is of little consequence to the conservationist whether one's initial interest in nonhuman nature is for pleasure or the advancement of science. What is relevant is that, if for no reason other than his own survival, man must soon adopt an *ethic* toward the environment. "The environment" encompasses all nonhuman elements in the one and only home we have on Earth. However, it will be some time before we are able to enunciate, much less promulgate, an environmental ethic because, fundamentally, the ethic runs contrary to our cultural tradition.

Ethics have been associated with man-to-man or man-to-society relationships. They have not been concerned with man's relationships to the nonhuman. Most moral philosophers have not acknowledged that man might have at least some ethical responsibility to the nonhuman. Perhaps this is because we cannot conceive of having any ethical responsibility to that which is not capable of reciprocating. Ethics, morals, fitness and propriety of behavior – these are human attributes. They do not exist, so far as we have been able to determine, in the nonhuman world. (That this may be a mere problem in communication does not seem to have occurred to us.) Since ethics do not exist in the nonhuman world, there is no need to apply them to that world. Our attitude toward the nonhuman world is not immoral: it is amoral.

In this book I concentrate on Western European cultures, not because those cultures have had any "corner" on ethics (they most conspicuously have not), but because Eastern ethics are becoming less and less relevant in the face of world industrialization in the Western image. Japan is probably the best current example of the abandonment of an ancient and nature-oriented ethos for the spirit of progress, Western-style.

The achievement of an ethic to guide our behavior toward the nonhuman world will inescapably involve the brutal violation of our cherished beliefs. It will necessitate close examination of our attitudes and assumptions, and of their origins.

Man's imagined removal from nature had its beginnings with the very dawn of culture. Whatever culture is, it is not biological. It is an artificial, conceptual inheritance, transmitted over the generations in the form of traditions, beliefs, revealed truths. Cultures change as they are passed on by the generations. They evolve at a much more rapid pace than do biological organisms. Therein lies our hope, and Earth's.

We cannot change our biological inheritance, but we can and do change our cultures – consciously. Conscious change of direction toward the environmental ethic will mean the practice of a kind of artificial selection – choosing certain positive elements in our traditions, and rejecting negative elements. The selective process will not be easy, for it will demand something that is foreign to us – humility. It will demand unprecedented humility in the face of the encountered facts of the biosphere and the cosmos. It will demand willingness to see ourselves in the perspective of time of infinite duration and of events of unimaginable magnitude.

CHAPTER TWO

One in a Million

The realization that I was "one in a million" burst on my consciousness years ago on encountering the astonishing fact that one human ejaculation may contain a quarter of a billion male sex cells. On later reflection, considering the potential number of ejaculations in a father's lifetime, the number of female sex cells produced, and the infinite possibilities of their combinations, I became keenly aware of "unique self." Even now my mind still boggles at the enormity of the odds against this peculiar individuality, and on the extraordinary nature of chance. Speculating on the outcome of even one sexual union, much less thousands, is as futile as attempting to count the stars in the night sky, to say nothing of the planets.

People tried to count stars long before it occurred to them to try counting sex cells. Both are fruitless; there are simply too many of them. Easier to reckon with are such phenomena as comets, which are often conspicuous and spectacular, and some of which are predictable. The most famous of these, Halley's Comet, has a cycle of 76 years, and if you miss it once, you are not likely to have a second chance. Halley's Comet last appeared in 1910. An old friend of mine, then at university, planned elaborately for a special viewing. On a particularly auspicious evening, he collected his girl friend, together with a blanket to sit on, and some refreshments, and repaired to the top of a nearby hill. It need not be added that, on that occasion at least, the comet was not seen. In his latter years, my friend (who by that time had become a distinguished ornithologist) would grin somewhat ruefully and shake his head, in full knowledge that he would not be around in 1986. He was fond of saying that some opportunities only arise once. And he was perfectly right, whether one's time-scale is human or cosmic.

The expression "Spaceship Earth" is in wide use nowadays, chiefly as a means of emphasizing the finite nature of our planet's life-support systems. On viewing Earth from a distance, as moon-voyagers have done, its isolation in the vastness of space is awesome. Neil Armstrong has remarked that he had not been involved in environmental issues until he observed Earth from afar, and recognized how relatively small – and how alone – it is.

Most astronomers estimate that our personal space vehicle has existed in its present form for about 4,500 million years. In other, embryonic and developing forms it has been around for considerably longer than that, but some element of controversy has surrounded its prenatal career.

Considering the mortality of species, faunas, planets, and solar systems, and the ephemeral nature of constellations and galaxies, the question of whether the universe has always been there or not is insignificant; it has little relevance to the transitory condition we on Earth call "real life." Real life on Earth will long be gone before there is any substantial change in what we now see in the heavens, and no one will be here to witness the outcome.

Our Earth spaceship is one of 100,000 million stars in one of the 1,000 million galaxies that lie within the range of our restricted observation. In the context of the one new solar system, the odds were weighted heavily against the emergence of life on one of a handful of planets, but in view of the number of suns and their satellites in our own galaxy – much less the universe – perhaps the appearance of living beings no longer seems remarkable.

Like all familiar living organisms, solar systems and planets have their beginnings, middles, and inevitably their ends. They are conceived, they develop, and they become extinct, both as individuals and as groups. It had to start somewhere, however, and it would appear that it began with the primal presence in the space-void of the most abundant element that exists – hydrogen.

In the immense black spaces between the planets and between the stars, there is little of anything, save a loose scattering of hydrogen atoms. They are not sufficient to form even a puff of vapor, for at this instant in cosmic time most of the supply of hydrogen is locked up in the sun and in the substance of the planets. Long before the birth of the sun, however, there seems to have been in this corner of our galaxy a certain concentration of hydrogen atoms which was sufficient to form a cloud of gas.

Here, the force of gravity came into play. Where the gas was sufficiently dense, gravity drew the component particles of the gas closer and closer together. The gas cloud began to form a ball. As mutual attraction between the particles increased, the ball became smaller, tighter, denser, until it shrank to less than one-millionth of its original diameter. The faster the particles of space-dust moved toward their center of attraction, the hotter they became. Eventually the center of the sphere became so hot (about 20 million degrees F.) that the atoms of hydrogen began to fuse. The original cloud of stellar gas had become a gigantic hydrogen bomb.

Though a relatively minor hydrogen bomb by cosmic standards, it is potent enough to generate the energy to sustain life on Earth. The thermonuclear process will continue until the sun runs out of hydrogen to fuse. As the supply of available hydrogen begins to dwindle, the sun will gradually begin to swell – at first imperceptibly, then more and more quickly. The planets will be consumed in orderly procession, beginning with the one nearest to the sun, Mercury. (Or, if there is a planet between Mercury and the sun, it will be the first.) Mercury will be followed by Venus, Venus by Earth, and so on. Scientists predict that after its swollen stage the sun will gradually shrink again, lose its heat and brightness, and disappear into the void from which it came. Observers in other galaxies will probably not notice its final twinkle.

The sun is now in the full warmth of its summer maturity – and so it must be to support life on any of its planets. The sun first acquired planets of its own because not all the hydrogen "dust" in local interstellar space went to feed the sun's thermonuclear plant. At a very early stage, the newly emerged sun was surrounded by a gigantic disc of galactic gas. The disc was far enough away to escape being drawn into the central inferno, but not so far as to escape the gravitational attraction of the sun. Gradually, these gases began to coalesce, in much the same way as the proto-sun had done, and to form the planets. Each gathered towards its nucleus whatever matter chanced to be drifting about, until there was little of anything left in what was now interplanetary space. Nine satellites had formed, and – *quite by chance* – one of them found itself at a critical distance from the sun.

The planet thus favored was the blue one. The Water planet. There happened to be atoms of oxygen in the original gaseous inventory, in sufficient number to combine with the ubiquitous hydrogen atoms.

Had the new planet been ever so slightly closer to the sun, its water would have remained in the form of vapor, due to the proximity of the sun's heat. Had it been ever so slightly farther away from the sun's warmth, its water would have remained in the form of ice crystals. No accident of fate was ever more significant to Earth than its critical distance from the sun, allowing for the thread-thin range of 100° C. within which the existence of liquid water is possible.

At the formative time of its "condensation," Earth was extremely cold, and its water was frozen. Robert Jastrow has described Earth at this stage of its gestation as "a Neopolitan sherbet of frozen water, ammonia and methane, plus various kinds of rocky substances, all immersed in a gaseous cloud of hydrogen and helium." However, a combination of factors which may have had to do with its own radio-activity, or with constant pounding by space-particles, resulted in Earth gradually warming. It warmed so much, in fact, that it melted. The ice crystals liquefied, boiled, and vaporized. Minerals, now molten, sorted themselves out according to their relative specific gravities, although the eventual distribution of what would be land masses was a long way in the future.

In the due passage of time, Earth began to cool off, starting naturally at its surface. The outermost layers of material congealed and solidified, and the cooling process slowly worked its way inward, leaving the molten core Earth retains today. But it was hot enough at the surface that the water supply was still in the form of vapor, which must have totally enveloped the evolving planet.

Since Earth's inception, the process had taken about 50 million years. Earth was a hot, but cooling, ball, surrounded by dense layers of clouds. But when the temperature finally dropped below the critical 100° C., the water vapor began to condense, and the rains started. How it must have rained! When we consider the enormous quantity of water now on Earth, the deluge from the skies must have been torrential. Presumably some of the rocks at Earth's surface still retained enough heat to vaporize much of the water, but eventually condensation overtook vaporization; and the result was our water cycle.

The lowering temperature caused Earth's surface to contract and to wrinkle. Unevenness at the surface resulted in the development of low catch-basins which soon filled with water running from the higher ridges in cataracts. Where it collected over wide areas, the embryo oceans developed. The sun, warming the surface of the seas, caused water to evaporate and to be carried into the primitive atmosphere, to be moved by weather systems around the globe until lowered temperatures caused it to condense and return to Earth as rain. Rain entered the rivers and thus the seas, and the hydrologic loop was complete.

Still, there was no life on Earth. The first atmosphere did not contain molecular oxygen – the product of green plants of the land and phytoplankton of the sea. Before these "producing" organisms, the atmosphere must have consisted of a mixture of water vapor and various other primal gases. And, as though waiting to be put to use, were immense quantities of liquid water at the surface of the planet.

The oldest recorded living things are organisms resembling bacteria, which have been found in Precambrian rocks in South Africa. These are fossils believed to be more than 3,000 million years old. The oldest fossils that are recognizable as plants are those of some algae found in Ontario; they are about 2,000 million years old. Other alga fossils, which may be somewhat older, have been found in the Rocky Mountains. These ancient plants are the earliest known organisms which were capable of photosynthesis, and they were probably the first sources of molecular oxygen and thus the originators of the secondary (breathable) atmosphere. Pouring the breath of life into the air, they made possible the emergence of the biosphere.

Where did these "pioneer" algae come from? Where did their progenitors originate? The basic "building blocks" of life are proteins, made of combinations of about twenty different kinds of amino acids. Scientists have synthesized some of these amino acids by firing electrical charges through such gases as ammonia, methane, water vapor, and hydrogen, all of which are thought to have been contained in the original, primary atmosphere. They were also in the original cold cloud from which Earth condensed. A spark of lightning may have begun the long "chain reaction." Ultra-violet radiation from the sun could have been a catalyst. By these or other means, or a combination of them, the ultimate experiment got under way.

Precipitation is our only source of fresh water. The original molecules of water were formed, it is thought, by the combination of hydrogen and oxygen atoms in the first cold cloud. The process may well have continued ever since, and the global water supply may be perpetually renewable from space. Some water molecules drift toward Earth from outside, but since most of the hydrogen atoms are components of the molecules of matter in general, and there are few of them loose in the atmosphere, it seems more likely that the water supply is finite, and whatever form it may be in at the moment – solid, liquid, or gas – the water we now have is all we ever had and may expect to have.

From oceans, lakes, streams and ponds, and from plant leaves and people, evaporation lifts about 100,000 cubic miles of water from Earth's surface for recirculation every year. Such is the force of solar power in the biosphere.

Although the water cycle is by far the most impressive of

Earth's life-support systems, there are others which are equally vital to the maintenance of living organisms. These are the carbon and nitrogen cycles, which make their unique contributions by way of the soil. Good soil is filled with multitudes of micro-organisms – among them, bacteria and fungi. These "plants" do not possess any of that miracle-stuff, chlorophyll. So they must operate on chemical energy instead of solar energy. These soil organisms are the power sources for the carbon and nitrogen cycles.

Green plants extract carbon dioxide from the ambient air for the manufacture of carbohydrates, releasing oxygen in the process. The carbon compounds thus retained in the structure of a plant get back into the soil when the plant withers and falls. Or the plant may be nipped off, before its time, by some grazing animal. In this case some of the carbon which was in the plant may be exhaled by the animal as carbon dioxide, and some of it may become part of the animal itself. In the latter eventuality, the carbon must await the demise of the animal before it can return to the soil. When that happens, the micro-organisms in the soil decompose the structure of the animal. The carbons enter into the bodies of these minute animals and plants, with an eventual return to the soil. Some of the carbons are oxidized, and thus get back into the air. The result is that the carbon component of living things is forever being cycled through the soil and into the air for subsequent use by plants.

Plants must also have nitrogen, but they cannot extract it from the air. So the nitrogen cycle consists of its being made available to plant roots in solution with other material, by way of the soil, again due to the contribution of micro-organisms in decomposition. An animal eats the plant, the animal dies, and the cycle is continued.

About three-quarters of the breathable atmosphere consists of nitrogen; the rest is mostly oxygen. The extraordinary thing about that part of the atmosphere which can support plants and animals is its thinness. Earth is about 8,000 miles in diameter and 25,000 miles in circumference, but there are only about seven and one-half miles of gaseous exchange above its surface. When puffing at the summit of Mount Everest (29,141 feet), or enjoying aperitifs in a jetliner at about that altitude – less than six miles – the air pressure is below one-third of that at sea level, and one cannot breathe without artificially supplied auxiliary oxygen.

Since the oxygen in the atmosphere is a biological product – the exhalations of green plants and of phytoplankton – everything ultimately depends on what goes on at the surface of the globe. The soil cover is, of course, much thinner than the breathable part of the atmosphere. The continents are said to be covered with soil to an average depth of from only seven to ten inches. Yet that micro-thin envelope has sustained itself and myriad living creatures for billions of years.

Earth, a relatively minor planet in terms of size, is by our standards extremely large, yet the biosphere is so thin as to be virtually transparent. It has been pointed out that if Earth were reduced to the size of a billiard ball, its mountainous surface would appear ivory smooth. We just might be able to recognize the great continents, but the oceans would be "mere films of dampness." We would not be able to detect the biosphere at all.

Slight and delicate though it is, the biosphere is sufficient for every life purpose. Miraculous? Perhaps not. Unique? Almost certainly not. In human terms, however, it is all there ever has been and all we can hope for. Earth – the product of sheer cosmic chance – may or may not be one in a million, but for our purposes it might just as well be.

CHAPTER THREE

The Concept of Community

In the lexicon of a group of scholars, the word "community" has a meaning vastly different from that given to it by a builder-developer – or by a sociologist, for that matter. Or by a global tribalist. To a biologist, community is something still different.

"Community," in the context of natural history, is a concept, not an absolute. The definition of any natural community is arbitrary and abstract for it depends always upon the scope in which one views it. For one man, observing the activity in a terrarium, that is a perfect community. For another, the community may be a pond; for another, a woodlot. For still another, the community may be the broad stretch of the Sargasso Sea.

The simplest community to *comprehend* is a lichen. There are many kinds of lichens throughout the world, most of them growing in such bleak and inhospitable surroundings that they may be the only sign of life. Lichens manage to eke out a living in some of the most biologically impoverished parts of the world chiefly because a lichen is not *a* plant, but a community. And a community is more than the sum of its parts.

Lichens are formed by a symbiotic relationship between two different kinds of plants, fungi and algae. The fungus contributes something and the alga contributes something; together they mutually benefit in what must be the world's most basic community. Lichens look vastly different, being of all colors and shapes, but each is a partnership in the very best sense. The fungus provides the framework or the foundation for the little cooperative enterprise. It secretes acids which help the lichen to maintain a foothold, and gradually draws mineral nutrients from beneath it. The alga, in turn, feeds the community. Algae have chlorophyll, the vital substance for the process of photosynthesis; fungi do not. Together, the two plants thrive as one plant; neither could exist without the other.

This interplay or interdependence between two forms of life in the same place at the same time constitutes a biotic community. It is simple in its number of components, but remains a real form of ecological cooperation.

Cooperation manifests itself on every scale throughout the natural world. One of the most remarkable cooperative communities that has ever evolved is one in which the aggregate constitutes what would appear to be one individual animal – the Portuguese man-of-war. This handsome "creature" is familiar to almost everyone who has beachcombed. Occasionally one will find dozens of small blue "balloons" scattered along the beach, where a group of the tiny sailors has been stranded by an onshore breeze. Although the blue (and sometimes pinkish-trimmed) bladder is the most conspicuous part of the man-of-war, attached to it there are slender, filament-like tentacles, which may be as much as fifty feet long. These tentacles are poisonous, and can deliver a fearsome sting. The man-of-war feeds on small marine organisms, which are knocked out or killed outright by the stinging tentacles, and are then digested in the "platform" which forms the lower part of the animal's float.

Like the individual lichen, the Portuguese man-of-war is not *an* organism, but a great many organisms living together in a cooperative community which supports all of them. There are at least four different kinds of animals involved, none of which could exist without the others. One animal is the float, or sail. Presented to the wind, it moves the man-of-war about the ocean. Another animal is the digestive apparatus. Still another looks after reproductive matters. A fourth is represented in the stinging and feeding tentacles, each one of which is an individual being. Far from being one creature, the Portuguese man-of-war is a community of animals, none of which could survive independently.

Lichens and men-of-war are composed of only a few kinds of participants. It is a short step, however, to communities of much greater complexity. A small terrarium may contain hundreds or even thousands (if we consider soil microfauna) of organisms, all dependent upon each other, all contributing to the success of the "undertaking" as a whole. The woodlot is another step up the scale of complexity, but the principle remains the same.

There are few free-loaders in natural communities. Some may give the impression of enjoying an unpaying ride at the expense of the societies with which they coexist; there are parasites on plants as well as on (and in) animals, and although they may live at the cost of their host, in the final act of its individual drama each of them contributes its remains to the nourishment of the community.

Some communities may appear more complex than would seem necessary. In a grassland community, for example, we expect to find primary "specialists" who carry out basic functions. There will be tillers of the soil, whether they be nematodes, earthworms, pocket gophers, ground squirrels, aardvarks, or elephants. There will be primary converters of plant energy, vegetarian grazers such as horses, cattle, antelopes, kangaroos. But in some favored parts of the world which have not yet succumbed to human pressures and where the natural way still prevails, we encounter more kinds of animals doing these specific jobs than we might have expected.

On an East African grassland there may be as many as a dozen different kinds of hoofed animals all living in a common area and all appearing to be involved in the same activity. Why this overlap of primary converters? On close observation, the realization emerges that although these different vegetarians *appear* to be doing the same thing in the same place, they are in fact not. The zebras are feeding in somewhat longer grass than most of the antelopes are. Wildebeest (gnus) are grazing on areas of somewhat shorter grass; it has already been "mowed" to the proper length (for gnus) by the zebras. Some of the little gazelles are grazing where the grass is even shorter, after the gnus have passed by. They are cropping the very smallest shoots. Each of them, specializing in its own way, is occupying a particular food niche, exploiting the resources of the environment in a slightly different way. The community has a place for everything – so long as everything stays in its place.

The system becomes misplaced only when two or more kinds of animals in a given community attempt to earn their living in precisely the same way. The structure of the community allows no room for such competition. Garrett Hardin puts it succinctly: *"Competitors cannot coexist."*

In nature, competition has none of the connotations of cut-and-thrust between individuals which human societies have permitted to evolve. Natural competition is an impersonal struggle for existence between species, without any consciousness of the struggle.

Unnatural or artificially induced competition is quite another matter. In parts of East Africa, chiefly that area generally known as Masailand, one can clearly see what happens when there is unnatural competition for a food niche in the grassland community. Domestic cattle have been widely introduced and are maintained in unfortunately large numbers. Their impact on the vegetation has been calamitous. Grazing at ground level, they concentrate on one food source – grass – building up unnatural pressure on the ground cover, which is soon eliminated. Erosion follows quickly. The impact of wild animals, on the other hand, which evolved not only with each other but also with the entire grassland community, is spread over the land both horizontally and vertically. Each specialist – from the toy-poodle-sized dik-dik to the eighteen-foot giraffe – is engaged in its own endeavor, and the pressure is never too great at any point in the food conversion process. Left to its own devices, the community functions smoothly. When a foreign ingredient is introduced, an unnatural form of competition results in gradual but certain breakdown.

There is more to a community than its biological component. There is the extra dimension represented by what Odum calls the "nonliving component" – the gases of the air, the minerals of the soil, and of course the water which sustains it. When we add this second major (nonbiological) component, what began as a community becomes an ecosystem.

Like a community, the size of an ecosystem is arbitrary. Some ecosystems are easy to define, as they are in the case of communities – a marsh, a forest, an island. However, it is impossible to describe an ecosystem in spatial terms; like the community, it is a concept rather than an absolute.

We live in an age of "systems." In essence, there are only two kinds of systems: open and closed. Almost all systems are open; they depend for their functioning on inputs from the outside. The terrarium or aquarium, considered as an ecosystem, is open; at least occasionally it needs air, water, or both, from the outside. The pond is an open system in that it depends on both water and air from outside. A spacecraft is an open system. In the absence of a recycling technology which would allow its occupants to live indefinitely on their own wastes (as inhabitants of the natural world have always done), the spacecraft must take essential supplies with it. The only closed ecosystem – functioning completely on its own – is the biosphere. It can be argued that the biosphere is not closed either: it depends upon the constant and uninterrupted supply of solar energy from the center of our planetary system. In a biological sense, however, Earth's systems are self-sustaining.

What each living thing depends upon is the constant flow of energy. Energy must pass through three fundamental stages or trophic levels. One consists of the green plants, which receive energy from the sun and fix it in the form of foodstuff for themselves and for all other living things. (This level also includes aquatic phytoplankton.) The second level consists of all animal life. No animal can fix solar energy; it must eat either plants or animals which have themselves eaten plants. Thus plants are producers of energy and animals are consumers of energy. The third trophic level consists of organisms called "reducers," the microscopic life in the soil – the fungi, bacteria, and so on – which break down the structures of dead plants and animals, releasing their components to the cycles of the biosphere.

The general trend of natural communities and ecosystems

is toward complexity. In the case of animals, we usually regard those which are more complex, or more "highly" organized, as the most sophisticated and successful products of evolution. In the same way, we see the more diverse and complexly interconnected ecosystem as the more stable. The more parts it has, the better its chances of remaining on an even keel for a longer period of time. A community such as a Portuguese man-of-war or a lichen can find itself obliterated should even one of its members be lost. A more complicated community, such as a woodlot, can, through shifting checks and balances, cope with almost any *natural* upset of one of its components.

Homeostasis, the state of equilibrium within a community, is achieved and maintained through constantly adjusting mechanisms within the structure which tend to compensate for fluctuations in any part, such as overproduction or overconsumption. There are constant shifts in emphasis and changes in numbers of plant and animal species. For example, cyclic increases and decreases in numbers of snowshoe hares are followed by corresponding fluctuations in lynx numbers. But the trend is always toward stability, with subtle adjustments constantly acting as though to keep the air bubble in the middle of a spirit level.

Since natural communities and ecosystems are continually moving toward multiplicity and diversity in order to keep the homeostatic level steady, the simplification of ecosystems is contradictory to every ground rule.

The simplification of landscapes is especially conspicuous in modern farming methods. In place of the scores of species of grasses which once were the base of the luxuriant prairie community, there is now a monoculture of wheat – one species of grass – which extends in every direction for hundreds of miles. Not only is this monoculture uninhabitable by most wildlife species, but it is also extremely vulnerable to attack by plant-eating insects and by plant diseases. Uninterrupted stands of grain – or of any other plant species – are open invitations to blights of every kind. Such bonanzas do not occur in nature.

Nowhere is there a better example of the simplification syndrome than in Canadian forestry operations. In the eyes of the forest industry, one of the most fearsome agents of destruction is the spruce budworm (which concentrates primarily on balsam fir). Its larvae eat the spring growth buds of the trees, and they have devastated great stretches of commercial forest. One of the reasons for the rapid spread of budworm infestations in recent times has been past cutting practices, and historic (one might almost say primeval) forest management techniques. Like the modern big-scale industrial farmer, the forest operator is interested only in certain species. To him, the best forest is the one that consists not only of a few commercially desirable species, but of stands of roughly the same age, so that cutting rotation may be carried out with greatest speed. The result has been enormous stands of balsam fir which, in much the same way as with wheat, encourage infestations of "pests."

In nature, trees of all ages would be intermixed with trees of many different species, thus lessening the danger of pestilence on the grand scale. To exacerbate the forestry problem, after a widespread budworm infestation there is the added danger of fire, due to the many dead trees. Even natural fire caused by lightning can thus be made more severe by human agency. Forest practice invited the budworm which prepared the way – the tinder – for the fire.

The simplification of natural systems, whether they be prairies or woodlands, is the inescapable result of the human drive toward specialization, in current times aided and abetted by economic forces. If he is in the business of growing wheat, the farmer grows all the wheat he possibly can, at the same time eliminating those plants he does not want, and which he calls "weeds." In so doing, he may well be removing the very buffer against depredation that his wheat crops need.

The commercial forester, interested in one tree or another, or a handful of species, attempts to eliminate "weed trees" in order to make his operations more economic – more competitive. The old-line traditional forester has been trained to examine the parts, not the whole. The holistic approach to forest ecology is unknown to him. Since "specialist thinking" pervades almost every aspect of human affairs, it comes as no surprise that comprehension of the totality of life has been slow to dawn on us.

Admittedly, one cannot "see" an ecosystem. No one stands back and says admiringly, "Now *there's* a lovely ecosystem!" But that should not mean we cannot comprehend it, or grasp the concept of community.

In the context of human affairs, we bandy the idea of community, but we are unable to see the human species in the larger scheme of the biological community. Man fails to acknowledge that he may be dependent on things other than human ingenuity.

In pondering the "community," we must allow our thoughts to range more widely than the human community – into the filigree of interrelated and interacting elements which form the system in which we live.

One may view a natural ecosystem in terms of a modern corporate phenomenon, the "conglomerate." A wide and diverse base of activities helps to ensure the stability of the enterprise as a whole. If the bottom falls out of SST's or petrochemicals, there are always hotel chains or distilleries. If one strand of the web weakens or snaps, there is always enough resilience in the structure to compensate for that loss, enough to swing into a new configuration of equilibrium, with the stresses distributed so evenly over all the other strands that no undue strain comes to bear at any point. Provided always that there is a sufficient variety of components.

The longer an ecosystem has existed, the greater the time that has been available for diversification and complexity, and thus for stability. Time is the essential ingredient for the development of stable communities. Obviously, the first rank in world ecosystems is occupied by the oceans. They have existed longer than any other communities, and they contain an incalculable volume and variety of living things. In addition to its plants and animals, the sea holds minerals and other materials delivered to it from the land masses in quantities we can only guess at. Its energy flows from phytoplankton (microscopic marine plants) to zooplankton (equally minute marine animals) through a bewildering array of lesser and greater invertebrates, fishes and mammals, through whom the cycle of life passes back to bottom-feeding decomposers on its return to the "great chain of being."

One of the ocean's most important contributions to the biosphere is oxygen. It has been estimated that as they go about their photosynthesizing activities, the phytoplankton release up to 70 per cent of the world's oxygen supply – besides being the fundamental base of the marine food pyramid. Recently it has been learned that even relatively small quantities of DDT, released into the sea from world river systems in concentrations of only a very few parts per billion, can reduce the photosynthetic capabilities of some kinds of marine phytoplankton. The conclusions are chilling. The oceans are vast, and in their antiquity (and thus their diversification and stability) they have been able to accommodate most of the insults which man has lavished upon them. That no longer appears to be the case. Byron declaimed that man's control stops at the shore, and that on the watery main all deeds were those of the ocean itself. The shadow of man's ravage is sickeningly evident today in the greatest of all ecosystems, from the sluggish shining oil slicks on its surface to the invisible but deadly synthetic molecules in its bottom mud.

The forests (especially in the tropics) offer the next widest variety of living beings in any given area. The typical equatorial forest consists of layers – from the great green canopy of the largest trees through the shade-tolerant forms which stand beneath them to lesser trees, shrubs, and smaller plants of the cool and luxuriant floor. Each stratum has its characteristic birds, insects, reptiles – even mammals. Such a forest is a welter and delight of sound and movement. One can almost hear the staccato heartbeats of those who live at such high-key intensity.

Forests are never static. Even with no human intervention, a forest is constantly in a state of transformation. Trees grow to their fullest extent, mature, and die. When at last they fall, the sunlight streaming through the gaps thus left in the canopy promotes the immediate growth of other species which theretofore had been inhibited by the shade. Later, fallen logs give both sustenance and substance to yet other species. New kinds of trees, new groups or associations of species are constantly "striving" for "succession."

Forest succession has been found to be reasonably predictable. After fire, flood, or clearing, it is possible to anticipate the kinds of small plants, shrubs, and eventually trees that will appear. They may not be the same species that were there before. The nature of the soil, exposure to wind and sunlight, and other factors all combine to determine who will succeed whom in the procession. In theory, the forest eventually reaches its "climax" – a steady-state beyond which it cannot go. The theory of climax must of course be viewed in perspective. Given a stationary state, with the influencing factors remaining constant, a forest will reach an "optimum" condition, dominated by trees best suited to the circumstances which predominate.

But things in nature do not remain constant, and the concept of climax must not be taken as an absolute. If North America is still in the midst of a warming period following the last (Wisconsin) glaciation, then the boreal forest of spruce, fir, and larch will, because of changes in climate, eventually move northward. To the south, it will be succeeded by the mixed and broadleaf forest in an orderly procession northward. On the western plains, it will be followed northward by the prairie grassland. The evergreen forest will "invade" the arctic tundra. *If* it is warming. If, on the other hand, we are about to enter a new glacial period, and the world climate should cool, then the reverse process would take place. These changes are of course imperceptible in the context of the human lifetime, but should we be present long enough to observe them, we would see a conspicuous process of succession over periods of thousands of years. "Climax," then, is useful in describing the dominant vegetation on a given site at a given point in time; but in reality – over the millennia – it is no more than an abstraction. For if the pure steady-state existed in nature, there could be no evolution.

On seeing an immense stand of forest, one often concludes that the soil must be rich and nutritious in order to produce such spectacular crops of greenstuff. In the tropics, unfortunately, this presupposition has led to many disastrous errors in land management. Generally speaking, tropical soils are not all that productive. In East Africa, the unglaciated soils have long since been leached of most of their nutrients by the annual rains. Where there are forests, the trees are shallow-rooted in the thin soil, and they survive because in the humid warmth the decomposition rate is high, and there is a fast turnover of nutrients. But when trees are cut, and the thin soil is exposed to sun, wind, and water, erosion occurs with shocking rapidity. Flying over mountains in South America and Africa, one is appalled at the devastation that follows removal of the forest cover. In many parts of the world, the substitution of agriculture for forest can only result in ecological and thus human hardship. Conditions prevail in natural situations for sound and established reasons, and we tinker with nature at our own peril.

Because they have existed for so long a time, oceans and forests are varied and stable ecosystems. At the other end of the scale are places that are surprisingly young – so young that they have only begun to realize their potential ecological diversity. Perhaps the best example of these, with the exception of volcanic islands in the ocean, is the arctic. Most of the arctic today is released from the grip of Ice Age glaciers for only a few short weeks each summer. In contrast to the forest, where innumerable generations of trees and shrubs have contributed their nutrient-filled remains to the food cycle, the arctic has virtually no soil at all. There has not yet been time for the weathering processes of nature to break down and distribute the rocky rubble left by the glaciers. Sometimes only inches beneath is the unrelenting permafrost, the permanent layer of year-round frozen earth, gravel and rock (or ice) into which water, and the roots of plants, can scarcely penetrate.

One major result of the permafrost and consequent poor drainage is rapid run-off of the spring and early summer meltwater. The formation of new soil is delayed for protracted lengths of time. These factors, together with the relative paucity of sunlight over the year, have limited the plant growth to tough sedges and grasses, and, of course, lichens. In addition, such "trees" as willows and alders scatter over the tundra north of tree-line proper, forsaking height for sturdy stems and ground-hugging form.

The tundra does, however, support a surprising number of animals. There are few species, but in some cases great numbers of individuals. Hordes of biting insects appear in summer, as far north as there is anything warm-blooded to feed upon. There are ground squirrels to dig wherever there is sufficient soil to dig in, and there are grizzlies to dig out the ground squirrels. Soggy tundras and deltas attract some of the world's most notable congregations of ducks, geese, and swans. And there are the unbelievable herds of caribou, which from the air remind one of the antelope which move annually over Tanzania's Serengeti Plain. No two parts of the world could be more different than the arctic tundra and the African veldt, yet unrelated species of animals behave in startlingly similar ways, in remarkably parallel adaptations to their respective habitats.

The solar energy budget of the arctic is so constrained that it requires vast stretches of landscape to sustain only a few animals. The herds of caribou (like the African wildebeest) cannot afford to be sedentary; if they were, they would overgraze and eliminate their own food supply. They keep moving, and their impact on the "reindeer" lichens which sustain them is spread thinly and evenly over vast stretches of landscape.

Arctic tundra ecosystems are critically vulnerable to

human disturbance. The Eskimos who lived in the arctic before the invasion by Europeans were hunters. Their impact on the homeostasis of the system, like that of most predators, was minimal. Today, however, with the advent of industrial man in the north, things may be expected to change at a rate heretofore unknown in nature. One of the gravest dangers to the arctic – paradoxical though it may seem – involves removal of the permafrost. Although its presence inhibits the development of luxuriant plant cover, the removal of the permafrost has devastating results.

When tracked vehicles move over the tundra in spring and summer, they sink into the softly thawed insulation of the ground cover, leaving trails. These tracks, open to twenty-four hours of daily sun, allow warmth to penetrate the denuded vegetative cover to thaw the permafrost. As the thawing proceeds, the tracks become wider and wider, runoff exposes still more permafrost to the sun, and the process continues to a point at which the track of one vehicle begins to resemble a muddy 1917 Flanders trench, though considerably longer. The eventual ecological implication of all this, which is occurring on a massive scale, is presently under investigation. One conspicuous result of rapid meltwater runoff thus induced is the simplification of the complex network of tundra tarns and ponds, which may be expected to have implications for nesting waterfowl.

Damage to the ground cover is becoming depressingly conspicuous in parts of the arctic now under exploration for fossil fuels. Aesthetically, the vehicle scars are deeply offensive, but it is difficult to put a price on unsightliness. Ecologists do not yet know the full extent of the damage to the tundra ecosystem, which is such an excruciatingly thin envelope over the all-pervasive permafrost. Perhaps before the arctic terrain is scarred beyond recognition by industrial invasion, and before its marine community also succumbs, there will be a sufficient body of scientific knowledge available to permit intelligent planning of northern exploration.

The arctic ecosystem is as yet relatively simple. This is

largely due to its youth. As time passes, a warming trend may bring new plants and animals to enrich the environment, to provide it with additional checks and balances, and thus with increased resistance to upset. A cooling trend may lock it all away, far from the reach of industrial man.

Through our failure to acknowledge our oneness with the total of other populations of biological organisms in our ecosystems, we reveal the egocentricity which has brought the world environment to the point of crisis. Reluctance to face facts which may be ego-destructive is self-deception. We have gone so far in our self-deception that the separation of man from his ecosystems has become even more than a deception. It has become an article of faith.

CHAPTER FOUR

The Numbers Game

Whether there is "purpose" in life is questionable, but there are definite trends. Biologically, the trend is toward diversity and variety, in order that the stable equilibrium of ecosystems may be maintained. Trends must involve a force, and the fundamental force in living organisms is reproduction. It would seem that the object of all living beings is to produce as many offspring as possible in order that a sufficient number may survive the exigencies of existence to replace those continuously being lost to natural attrition. In some parts of the world, men want to have many sons to ensure that there will be someone to care for them in their later years. In nature, the attempt is to deal with the overwhelming odds against the survival of any individual or group of offspring.

A population consists of all individuals of one species in a given community. In studying natural populations, there are a number of elements involved. How *dense* is the population: how many individuals of that kind of organism are there in the area being observed? In the case of some invertebrates, one might be counting numbers per square foot; in the case of redwinged blackbirds, numbers per acre; in the case of deer, numbers per square mile.

The density of a population of animals is partly a reflection of the species' birth rate. Animals and plants have spectacularly different birth rates, according to their ways of life and their roles in the community. Plants, which are the primary food producers and are eventually devoured by everyone in the community (including each other), must perforce have a high rate of reproduction. Vegetarian animals in general have higher birth rates than meat-eating animals, because as primary converters they too are suppliers to the food pyramid. Their "turnover" is high. To understand what is happening to animal populations we must know not only the birth rate of a species but also its rate of turnover, and this involves its death rate.

Assessing populations is difficult. Many organisms are anything but sedentary. They come and go. Populations in local areas do not remain static. Nothing in nature does. There are individuals who, for various reasons, move away into new communities, and there are those who wander into an area from outside.

Despite fluctuations, natural populations (with such exceptional phenomena as lemming cycles) give the *appearance* of being relatively stable. Birdwatchers notice minor or even major changes in some species over periods of several years, but they also learn to expect reasonably dependable relative populations of the most familiar species. It was once thought that some mysterious mechanism was built into a species which prevented its numbers from "exploding"; somehow the birth and death rate were counterbalanced. However, this is not the case. Changes in the environment of a species may have extraordinary consequences; one change might wipe out an animal completely, while another might allow it to increase at an incredible rate. This is precisely what has happened with the human species, and it has been observed in many others.

There is a valid analogy between the uncontrolled growth of human populations and the unnatural increase of non-human species which man has introduced into new environments. Man has manipulated his own habitat to the point of creating a new kind of environment expressly suited to his wishes, and this new and special set of conditions has involved the elimination of most natural population controls. The same phenomenon occurs when we introduce a bird or other animal into a set of environmental conditions which is foreign to it. In most cases, the new species fails to survive because it is not adapted to the exotic situation. Occasionally, a new species may find itself able to cope with foreign conditions, and free from the usual biological controls in its original habitat. In such cases, the result can be impressive.

The deliberate introduction of the European starling and the house sparrow to North America, and of the rabbit to Australia, shows the consequences to animal populations in the absence of homeostatic controls. Virtually overnight, the birds and the rabbits increased to "pest" proportions in their new freedom from natural restraints. That is because the birth rate of most animals is as high as local conditions will allow. Among many small invertebrates and fishes the odds against the survival of an individual egg are of the order of thousands to one. In the spores of fungi, the numbers may be astronomical.

In one breeding season, most birds "attempt" to raise as many young as the traffic will bear – or as many as is physiologically possible for them. The major limitation on bird increase is the availability of food for their young. The constraints exerted by food supply cannot be demonstrated better than by the hawks and owls. Most birds do not begin to incubate their eggs until the entire clutch has been laid, and thus all the young hatch at the same time. But with birds of prey, all eggs are incubated from the moment they are laid. Since eggs are produced on successive days, or at longer intervals, the brood will consist of young hatchlings of varying ages and sizes. In times of plenty, all will survive. In meager times, only the strongest nestlings (usually the first-hatched) will make it. When food shortage becomes acute, only the largest chick will come through, often by devouring its nestmates. This is a form of homeostatic population/food adjustment which operates according to the environmental pressures (or the opportunities) of the moment.

Smaller birds produce all the young of which they are capable and "hope for the best." The most familiar example among the songbirds is the backyard American robin. In the latitude where I live, robins generally nest twice in the course of a spring and summer, with a normal complement of four eggs to each clutch. If all of these hatched, and if all the young were to survive, we would have eight first-year birds and two old birds back in the garden next spring. Ten where there had been two. Games of arithmetic do not prove much, for exponential curves exist only on paper, not in nature. In theory, however, if robin reproductive success were to continue at this rate without interruption, with adults and young surviving for ten years (not an unreasonable age for a lucky robin), there would be 19,000,000 robins in the garden at the end of that fateful decade. And there would be an equal number in the garden of my neighbor. But this of course does not happen. Next spring there will again be two robins in my garden, not ten. Like it or not, God *must* see the songbird fall. In fact, He has to see to it that something of the order of 80 per cent of the robins are dealt with in a normal year.

There are abnormal years and abnormal series of years. In the middle and late 1950's, when the impact of DDT and other chlorinated hydrocarbon insecticides was beginning to be widely noticed, there was a notable die-off of robins. DDT sprayed on elm trees, in attempts to control elm bark beetles and thus the imported Dutch elm disease, was carried to the ground on leaves in the autumn. There it was transferred to earthworms which ate the leaves, thence to robins. George J. Wallace, studying the robins on the campus of Michigan State University, estimated that ten such earthworms, loaded with DDT, were sufficient to kill a robin. He reported that in only four years his local robin population on campus dropped from 370 to three. During the same period, robin populations in general were seen to dwindle over substantial areas.

The robins eventually picked up again, and robin populations seem to be relatively stable once more. But in the interval – after the bottom had dropped out and the long climb back had begun – somewhat more than 20 per cent of the robin population was allowed by the environment to survive in a one-year period.

Under natural and undisturbed conditions, there are a number of factors which tend to limit outbreaks of the proportions we have experienced with house sparrows and starlings, and which we could in theory experience with robins. One of these is crowding. Some large birds, such as eagles, resist the company of their kind (for reasons having to do with foraging space), to the extent of being entirely solitary in their nesting habits. Other birds seem to "demand" the most intimate togetherness. Great colonies of auks, such as murres, and of penguins and other seabirds are common. Some bird species will not come into breeding condition, or will not breed, without the stimulation of a dense surrounding colony of their kind.

A point occurs beyond which these seabird colonies do not appear to grow – as though an invisible dome were placed over the colony, allowing no further expansion. Certain gannetries in the world are known to have increased substantially over a period of years, but this is probably the result of protection. With the pressure off, the colonies have reverted to their "normal" size, where they will level off again.

In colonies of albatrosses, gannets, or auks such as puffins, the "core" of the colony usually consists of mature birds which have bred successfully in prior years. Upon arriving at the colony at the outset of the breeding season, these older and more experienced birds take up the choice nesting sites in the center of the area. There, they are less vulnerable to predation, and, perhaps, more stimulated by the influences of sociality. The younger, inexperienced birds take up positions around the periphery of the colony, where they are more exposed to accident, and where they may not realize the benefits of a pleasant degree of crowding.

These younger birds may or may not be paired. They may go through the motions of display and preliminary site-choosing, and they may not breed at all. (In some albatrosses, the birds are not sexually mature for six or seven years, and even then they may not nest successfully the first year or two.) The young birds go through this ritual "practice," or they loaf about the outskirts of the colony, not even going through the most perfunctory nesting ritual. They become observers, and they may observe more than we surmise. In some kinds of birds, however, breeding is possible even when one or both members of a pair are still in sub-adult plumage. In a disaster year, when for one reason or another a large proportion of the fully adult birds may be removed from the colony, these younger individuals may be recruited into the main body of breeders. In such cases, they may nest successfully, as though to quickly bring the population density back to the proper level. They represent a pool from which the species may draw in times of necessity.

The pooling of potential breeders is nowhere more evident than in colonies of fur seals and sea lions. Here, a relatively few fully adult males hold all of the females in harems which are more or less large. The younger or older or incapacitated bulls form bachelor herds of their own. At the first sign of weakness or vulnerability on the part of one of the dominant beachmasters, there is always another bull ready to do him battle and to drive him away and take his place. This, however, has nothing to do with population *density* in the strict sense, for all the females in the colony are impregnated, regardless of which bull may have done the impregnating.

Without evidence that fluctuations in birth rate contribute to the stability of bird populations, control must be exercised by the death rate. Death rates change conspicuously. These are probably highest when bird densities are at their peak (when the auk rookery is "full"), and at their minimum when densities are lowest. The most prominent causes of increased death rates when animal populations are high are disease, predation, and food constraints. Various diseases are known among mammals. On one visit to the Galapagos Islands I noticed that the local sea lions were more numerous than they had been six years before. I noticed also that the beaches were littered with dead sea lions. A mysterious "blister disease" was sweeping through the colonies, killing old and young alike. On the basis of unusually high sea lion numbers, one may speculate that the disease is a response to population density. One may also anticipate a new equilibrium in due course.

More familiar mammal diseases – especially those of the wild dogs such as wolves, coyotes and foxes – are distemper and rabies. It does not appear certain that these are devices of population control, for numbers of predatory animals are stringently controlled by the relative numbers of the animals they eat. The diseases are probably always present in population "reservoirs"; and in times of malnutrition or other adverse conditions, they may tend to eliminate a certain number of young animals and those approaching senility whose defenses are lower. There are of course sporadic wide-scale outbreaks of both rabies and distemper, in which larger than usual numbers of animals may be afflicted, but these do not *seem* to be related to population regulation. Among all animals, disease generally comes into play as a controlling agent when there is a substantial "surplus" of a particular species. This surplus, because of lack of food or other factors, is more prone to illness.

Predation appears to be in much the same category as disease. Under stable conditions, predation generally strikes the young and the aged – the weak and the infirm. It is rare for any predator to significantly reduce its own food supply. The survival of predators may be larger than average in years of unusual food abundance and it may drop when the pinch is on. There is a clear relationship between the number of predators and the number of animals available to them for food. The controlling agent in the predator-prey relationship is the prey species rather than the predator. Again, this is a homeostatic control. As. V. C. Wynne-Edwards remarks with regard to senile mortality, it "occurs at the age that suits the species concerned, more or less independently of the particular agency that happens to discharge the death sentence."

Discharge of the death sentence *must* be carried out. Among animal species there is no avoidance of population limitation. A provocative agent of population control is Hans Selye's "general adaptation syndrome," relating to stress. This works chiefly on adults in a confined area.

Desmond Morris, whose books include *The Naked Ape* and *The Human Zoo*, is an authority on captive animals. No zoo-keeper in his right mind, Morris feels, would dream of confining animals in such restricted quarters as human city-dwellers find themselves. Overcrowding in animals results in increased aggressiveness toward others of the same species, in quarreling, and sometimes in death. This mode of behavior will keep population increase in line. But it may go beyond that, to cut down on absolute numbers.

Stress need not build to the point of blood-letting. As Selye has shown, many changes may be occurring within the individual in a physiological way which may not at first manifest themselves externally. These changes may show up in the adrenal glands, the thymus, spleen, and sundry other lymphatic tissues, the red cells in the blood, the mammary glands, the reproductive system, and the fat reserves.

The most interesting homeostatic control of natural populations is the phenomenon of territoriality. Many kinds of animals show a tendency to appropriate a certain physical area of their environment and to defend it against others of the same species. They may do this as individuals, as breeding pairs, or as groups or bands. The practice is more conspicuous among birds than mammals, although it is also characteristic of certain fishes and reptiles.

Territoriality appears to be a response to fluctuating food supply. Wynne-Edwards believes that one of the effects of territoriality is the spacing out of individuals of a population of animals, and that the practice not only produces dispersal of the animals but also functions to control population size. John Hurrell Crook states that "this hypothesis is part of a broader theory which holds that social organization is essentially a mechanism for providing 'conventional competition' whereby numbers are regulated by homeostatic dispersal in relation to food resources." Conventional competition avoids physical combat which would otherwise erupt under the stress of overcrowding.

It was H. Eliot Howard, a British student of birds, who in 1920 revealed to the world the widespread prevalence of territories among small songbirds. Since that time, a growing number of people have been led to study territoriality, but knowledge of it – in its full complexity – is far from complete. There is little outward consistency about territoriality. In East Africa, spotted hyenas and Cape hunting dogs get their living in essentially the same way – by pursuing and pulling down such prey as antelopes. The hyenas are territorial, while the dogs are not. One must probe deeply to ferret out just what it is in the life style of an animal that determines whether that species will or will not be territorial. Sometimes the same species may behave differently in different parts of its range.

There are vast differences in the size of animal territories. They may consist of many square miles, as in the case of a pride of lions, or they may cover only a few square inches, as in the case of a murre or other seabird virtually rubbing shoulders with its neighbor. The lions' hunting territory must be large enough to maintain a stock of game, or at least to allow game to pass through. The murres' fish-hunting area is the open sea. The only terrestrial real estate a murre needs is adequate space within which to incubate its single egg.

Some animals maintain territories only during the breeding season, some the year round. Others hold different territories in summer and winter. Most of our familiar migratory songbirds in temperate zones, such as thrushes and warblers, are vigorous in the defense of their nesting territories during the summer, but during the off-season they congregate in close-knit flocks, with no sign of the territorial jealousy they exhibited only a few weeks before. Whooping cranes, on the other hand, have breeding territories in northern Canada during the summer, and off-season territories on the Gulf coast of Texas in the winter. European robins, which are not migratory, have both summer and winter territories.

The defense of territory is often confused with aggressiveness. Such defense is merely the response of the individual animal to the intrusion of a stranger of the same species. As someone once said, ironically, "The bear is dangerous: when attacked, it defends itself."

Defense of territory takes many forms. Most small birds define the limits of their plots by means of song; they have favorite singing perches which are situated about the territory and which distinctly mark its boundaries. Many mammals, such as dogs and cats, leave scent markings to indicate their proprietorship. In addition, as though to reinforce their scent markings, wolves will howl and lions will roar. Gibbons whoop noisily, and howler monkeys roar and growl. In the case of wolves, howling serves in most instances to keep packs from ever meeting up with each other. Thus, the population of a given species spreads smoothly through favorable habitat. Territoriality probably keeps the populations of a species forever exploring new areas, thus pressing toward the ultimate limits of suitable environment.

Man has been emancipated from most of the controls which curtail the increase of populations in other species of animals. We are not bothered any longer by predators – at least most of us are not. Western man has few food problems, and relatively few from contagious diseases. We do not practice infanticide, nor are we cannibals. Most of us have even survived the effects of stress. We have escaped most of the constraints to which other animal species are inextricably bound. The result has been an unnatural increase in human numbers, to the point where we endanger not only ourselves but the systems which give us life. By the sheer weight of our numbers, we threaten that magnificent kaleidoscope of other living things which also depend on the natural systems.

One of the most instructive accounts of the growth of human numbers is Goran Ohlin's background paper for the 1965 United Nations World Population Conference, *Historical Outline of World Population Growth*. It has served as the basis for most of the following estimates.

During his earlier days as a Pleistocene hunter and gatherer, *Homo sapiens* was a reasonably well-integrated member of the natural community. At the beginning, he was probably little more efficient than most social predators (which is to say, no more than middling-efficient), and he must have had only minor impact on the numbers of the animals he hunted. But his hunting improved, as the large mammals of the late Pleistocene learned to their cost, and he began to superimpose a technological tradition upon his earlier culture. Human populations at this time may have been as few as one million, or as great – at the outside – as ten million. We shall never know precisely how many men there were, but we can be sure that they represented a drop in the bucket relative to the total biomass of the great Age of Mammals. It was not until somewhat later that man began to have an obvious effect on the landscape. The destruction of the natural world began with civilization.

If we date civilization from the "agricultural revolution," that brings us to approximately 10,000 years ago. The partial domestication of certain animals presumably came before and may have led to the domestication of plants. Even more important to the environment was the end of the age of nomadism. It was the end of man's natural integration with the land and with the biotic community.

Because they kept moving, wandering pastoralists must have had only slightly more impact on the land than did the bison. (Goats and sheep, however, can stay in one place until they eat it down to the bare earth, and no doubt this happened quite early. But cattle are not as deadly because they are less efficient at reducing grass to desert, and periodically they must be moved to greener pastures.) But as soon as man stopped wandering and began to tend crops, permanent communities were formed, and this was the beginning of widespread environmental degradation in the Mediterranean basin. From an absolute maximum of perhaps 10 million during Pleistocene pre-history, the numbers of people jumped to at least 50 million by classical times. Communities had also become groups of communities, super-communities had become city-states, which in turn had become nations and empires.

By the time of Christ, the world population had grown to something between 200 and 300 million. Bubonic plague and other epidemics reduced the population; but by 1650, when Europeans were busy with global exploration and conquest, the number had reached about 500 million.

In another hundred years, while Jean-Jacques Rousseau was yearning for a return to the Arcadia of our forebears, the world human population had reached 750 million. It is no wonder that a romantic misanthropy was pervading Europe at the time. By 1850, due largely to advances in sanitation, public health, and medicine, we had reached our first billion. Headlong expansion in the New World had followed the ages of discovery and colonization, and the industrial revolution; and in another hundred years, by 1950, the total had reached two and one-half billion. We have done rather well since, due to such agents as DDT. Today we stand at approximately 3.7 billion – a rise of over one billion in what used to be considered the standard human generation. By 1980 it could easily be 6 billion.

The absolute numbers of people are striking enough, but the *trends* are even more impressive. Before the agricultural revolution, it is believed that the human population could double only in from 10,000 to 100,000 years. In the period between the agricultural revolution and the time of Christ the population doubled every 1,000 years. Between the year one and 1650 there were the effects of the Black Death, and the population doubled only at the rate of once every 2,000 years. But it picked up quickly, and doubled again in only 200 years between 1650 and 1850. It doubled again in the next 100 years. At present, the estimated doubling time is between 35 and 40 years. Yet there remain in our midst an astonishing number of people who would apply standards and criteria to human reproductive behavior which may not have been appropriate even in the Dark Ages.

One of the greatest ironies is our humanitarianism. I have held in my arms a Guatemalan Indian child dying of malnutrition, and such is the nature of our species that I would have done anything in my power had I been able to save that baby. Many are being saved. Control of communicable disease, improvement of sanitation, control of infant mortality, massive advances in geriatrics, have all resulted in "death control." Pasteur, Jenner, Lister, Fleming, Banting, have helped alleviate the general human condition. Our population almost doubled in the short interval between Paul Müller's award of the

Nobel Prize for his discovery of the insecticidal properties of DDT in 1939, and the 1962 publication of Rachel Carson's *Silent Spring*.

It is impossible to be indifferent to the story of DDT. It *is* possible to be ambivalent about it. I am distressed by the effect of DDT on the bald eagle, the osprey, the peregrine falcon, and the brown pelican; but I was also distressed when a malarial mosquito in Africa's Rift Valley assaulted me. Had I been weak or undernourished, the mosquito could have killed me, or so debilitated me that some other disease – perhaps the flu – could have finished the job. No doubt the mosquito *should* have killed me, because there are too many of my species. There is no doubt that that is the malarial mosquito's function. Whatever the consequences, DDT has saved innumerable people from insect-borne diseases. This has resulted in many more people.

Vasectomy – male sterilization – is suggested as the best contraceptive device. It is not very encouraging, however, on the basis of a little mental arithmetic. Generally, the operation takes about half an hour. An eight-hour day, with an hour off for lunch, represents fourteen vascectomies per surgeon per day. A team of three surgeons, working around the clock on eight-hour shifts, could perform 42 operations per day. Let us imagine further a clinic with ten operating tables and thirty surgeons, and our count comes to 420 sterilizations per day. Assuming no holidays or days off for this dedicated team, we come up with 153,300 operations per annum. How many clinics of thirty surgeons would be needed to take care of even one-tenth of the world's estimated 1,850,000,000 males? To say nothing of an equal number of males projected for the next 35 years.

Man has already exceeded the "carrying capacity" of planet Earth for his species, and in so doing he has crushed the life from species that, in the Pleistocene, were his brothers. There has been a breakdown in homeostasis. Despite the brash confidence of those who promote agricultural technology and the "Green Revolution," and despite the best intentions of those who call for the redistribution of wealth and food, nothing is working.

In his historic address to the United Nations in 1965, Pope Paul said (in translation), "Your task is to ensure that bread is sufficiently abundant on the table of humanity, and not to favor artificial control of births, which would be irrational, in order to diminish the number of guests at the banquet of life." Life in underprivileged countries is no banquet, His Holiness notwithstanding. Also, no banquet was ever crashed by a babe in arms: it is seared into my memory that the dying Guatemalan infant did not ask to be there.

CHAPTER FIVE

Bald Eagles and Pot-Bellied Pigs

If, as would seem to be the case, modern man has existed in recognizable form for about a million years, it means that that particular evolutionary experiment has been going on for about one three-thousandth of the history of life on Earth, and for about one four-thousand-five-hundredth of the history of Earth itself.

Richard Carrington has used the device of compressing time scales to help our perspective. Let us imagine the entire history of planet Earth as speeded up so as to occupy one 365-day year. For the first four months, there is no life. There are no birds or mammals until mid-December. Shortly after ten o'clock on New Year's Eve, there is proto-man. Recorded human history would be lost in the initial whir of the clock striking midnight. Such has been the time for strutting and fretting allotted to the poor player by the master script. Shakespeare, in one of his numerous misanthropic moments, concluded that one lifetime of sound and fury signified nothing. No doubt, had he known biology, he would have said the same for the lifetime of a species.

Sometimes it is useful to think of single organisms as groups of organisms, and vice versa. A human being is made up of millions upon millions of individual cells, so a human being is a community of cells. A termite hill, comprised of thousands of individual termites, can be thought of as one animal with its component cells. Cells are forever dying and being replaced – as are individuals and species. Yet many persons continue to maintain "the immortality of man."

Like a star, or a planet, a species is an individual being in that it is born, it lives, and it disappears. I am not an exact copy of my father, or my grandfather; and my sons and I are not identical. I evolved from other men who looked different from me, and other men evolve from me, but not in my image. Modern man evolved from an earlier kind of man who also looked different; and the lifetime of both, as species, is limited.

My personal extinction comes with my death; and since people make it their business to care about such things, the date and even the hour of my leave-taking may well be known. But the lifetime of a species is so long that its death (as well as its birth) is usually blurred, and it is often very difficult to determine the precise point at which it disappeared. That is because there are two ways of becoming extinct.

We know the date (September 1, 1914) of the death of the last passenger pigeon, for the species was wiped out abruptly and the last individual perished in a cage. We shall never know, however, the date of the passing of the proto-passenger pigeon, for it evolved into something else. In one sense it never became extinct; it lived on in the passenger pigeon which it had become. On the other hand, one could claim that the proto-passenger pigeon became extinct when the new passenger pigeon became an entity. Whether *Homo erectus* still lives, in slightly revised form, we do not yet know; but we are fairly sure that a kind of *Australopithecus* still exists, albeit in substantially altered form.

The same considerations apply to any species, present or extinct. I was there – and so were you – and so were the dinosaur and the sabre-tooth and the Neanderthal, when Jehovah released the first lightning bolt into the primal broth. It has been a long continuum, and it is not yet over. Some of us are still here physically, and all of us always were here genetically. And those of us who are here now – hummingbirds and dandelions and people and squids and damsel flies – have an unbroken genealogy that goes back to the amino acid soup. We were all in it together. We still are.

The fossil record of past ages of life is essentially a history of extinctions, for none of the ancient creatures are alive today – at least, not in the same form. The fossil record is also the history of an uncountable number of experiments. Each new living thing – each individual plant or animal – is an experiment. Each is an "attempt" to probe farther into the possibilities of the biosphere. Like human fingerprints or zebra stripes or giraffe blotches, no two genetic patterns are identical. Each new individual differs from its progenitor.

By mixing the genetic inheritances of two individuals, sexual reproduction creates a unique being. Random genetic fluctuations in the offspring further ensure that change – the one constant in all of life – continues. Garrett Hardin describes heredity as "fate's lottery," and the way in which chance enters into the fate of each new individual as "The Great Fertilization Sweepstakes."

Many persons find the element of chance in life distasteful, even repellent. However, appreciation of the enormous odds involved in the lottery which is human conception should make us even more aware of our individual uniqueness. Chance is the element which keeps everything else going, whether we like to admit it or not.

Genetic change is in fact change for the sake of change. Animal and plant species must change in order to survive in changing environments. Over the history of the planet there have been constant climatic and geological transformations. Wet hot periods and dry hot periods, frozen periods and temperate periods, have followed upon each other since the world began. There have been times of mountain-building, flooding, and continental drift.

In the course of this never-ending flux, plants and animals of various kinds have encountered a bewildering variety of circumstances. Reptiles can tolerate neither intense heat nor intense cold, because of their inability to regulate their internal temperature, and the peaks and hollows in reptile fortunes have been conspicuous. Mammals and birds, on the other hand, with their homeostatic warm blood control, can venture anywhere in any kind of weather. They can survive from the equator to the pole, so long as there is food and shelter. Over the long haul, birds and mammals have been more "successful" than reptiles. They have penetrated to every corner of the globe.

Evolution involves perpetual exploration of new environmental opportunities. Life as a whole may be described as "opportunistic." As in reproduction, the tendency is to press against the extreme limits of possibility. One might picture a lichen, or perhaps some small vine, as representative of life. With its newest shoots or tendrils it is forever exploring new cracks, new crevices and niches in which it may fasten the tip of a searching growth-bud and there perpetuate itself.

Birds provide ready examples of the opportunistic exploratory process. From the nucleic bird, whatever it may have been like (something had to precede the feathered *Archaeopteryx*), a million different kinds of birds evolved, of which some 8,600 are on Earth at the moment. Each of these species represents the green searching tip of a tendril, always in readiness for new opportunities.

The growth tip of one of the 8,600 bird tendrils is the green heron. There is in South America a very similar and closely related species, the streaked heron, representing the other half of the last fork in that particular tendril. Now, however, the green heron shoot has turned another unexpected corner and encountered a new and empty niche, and it is no longer a green heron. It is sooty black. It has found itself in the Galapagos Islands, and there it has evolved into something new – a lava heron. The green heron on the mainland has remained the same, but it has produced a new species before our very eyes.

The Galapagos, an archipelago in the tropical Pacific, has made an inestimable contribution to our understanding of the processes of evolution. Oceanic islands, removed from the mainstream of life on large continental land masses, have often gone their own peculiar ways. The Hawaiian islands of the Pacific, the Mascarenes, Seychelles and Aldabra of the Indian Ocean, and many others, have all produced unique faunas. But Galapagos is supreme as the classic instance of the appearance of strange endemic wildlife populations. Galapagos iguanas, tortoises, lava lizards, gulls, and finches, are world-famous.

Isolation from continental biological influences allows oceanic islands to give rise to new and unique beings. In the case of Galapagos, which lies on the equator about 600 miles west of the coast of Ecuador, the distance is sufficient to prevent many of the more sedentary forms of life from intermingling with its inhabitants, and this has hastened the development of a new being from something very old.

The beauty of Galapagos is not its age, but its *youth.* Some of the islands are no older than the genus *Homo,* which is very young. By comparison with the South American mainland, Galapagos was born yesterday.

The islands are the individual summits of immense submarine volcanoes which have risen from the floor of the sea some two miles below. Current understanding of geophysical phenomena would indicate that these volcanoes (some of which are still highly active) are manifestations of ocean bed movements associated with the drift of continents.

Once risen from the sea, the islands were nothing but bleak and barren cones of hardened lava. The rocks were bare and lifeless. Eventually, prevailing winds brought the minute spores of pioneering lichens. Due to their fungus component, the lichens were able to take a foothold, and to spread. (The islands today are a lichenologist's paradise. One expert spent a full season on one island, and such was the profusion and variety of his favorite plants that he was able to work his way only a quarter of a mile inland.)

The work of lichens is remarkably slow, but they have ample time and they are the hardiest of plants. They may live for thousands of years. Gradually they break down their rocky bases into minute soil fragments which over the years accumulate in cracks and crevices. Thus the way is prepared for the adventitious arrival on the wind of small-seeded plants such as grasses. Erosion helps the process, and eventually the volcanic rock is broken down into a thin and still tentative covering of soil, in which larger and more elaborate plants can become established. Also, the presence of grasses and other plants may then make it possible for plant-eating animals, such as insects, to survive if they happen to be swept on some weather system to the islands.

At any given moment there must be tremendous numbers of insects and other small invertebrates being carried from place to place on the winds. Most of those which find themselves over the oceans will perish. A few, however, might be deposited on some oceanic island. If they find no suitable food, they too will die.

If the insect happens to arrive after its favorite food plant has become established, then it may well survive. The environment has allowed it to pass through the "selective screen." This, it appears, is the way animals first became established on islands such as Galapagos, and it illustrates the screening or filtering process which Darwin described as "natural selection."

The same mechanism operates on larger and more complicated kinds of animals, whose needs may be just as specialized. There are two species of flycatchers in Galapagos. A wandering, wind-buffeted flycatcher could not have survived upon arrival in the islands if there had not already been an established population of flying insects to sustain it. In the flycatchers' case, the selective screen was represented by the presence or absence of flies. Other animals, such as lizards and snakes, which could not be carried on the winds, no doubt arrived in Galapagos on rafts of flotsam which drifted on prevailing currents from the mainland. In the case of iguanas, the selective screen would be the presence or absence of certain kinds of vegetation, such as cactus. The snake would need a supply of small animals. One way or another, there was an orderly sequence of events which allowed the islands to be invaded and subsequently populated by gliders, fliers, drifters and seafarers from the Americas.

It is probable that only a few individuals of each species arrived – perhaps half a dozen tropical mockingbirds, one pregnant bat, one durable egg-case of an invertebrate without any adult being at all. But the small number of invaders is significant in an evolutionary sense. Those that survived long enough to reproduce passed on their peculiar characteristics. There were so few of them that Galapagos wildlife populations began with a heavily biased sample of the genetic potential of the species concerned. Sheer chance had brought certain individuals, and since each individual is genetically different from all others of a species, the invaders represented peculiarities not common to their species as a whole.

Animals of all kinds have a tendency to throw off "sports" or "mutants" from time to time. In a brood of young animals, there may be one which lacks pigment – a white one, or albino. Usually these do not survive for very long. They are either so conspicuous that they represent an easy target for predators, or they may not be accepted into breeding populations of their species, and be screened out socially. Even if they find mates, their albinism is usually "bred out" in subsequent generations.

In the case of melanism, an animal receives a larger quantity of dark pigment in its inheritance. The black panther is a melanistic leopard; the black squirrel is a melanistic version of the gray squirrel. In many species, these dark "sports" survive to lead normal lives. Their black color may, in fact, serve as camouflage. It matters little what color a night-hunting leopard may be. For the little heron in Galapagos, the black lava flows might serve as excellent concealment. The heron a little darker than the others might live longer, breed in more successive seasons, and thus have more survivors. Most important, it would pass on the darkness which allowed it to live longer. Gradually, the number of dark herons would become greater and greater, and the number of more highly colored individuals would drop proportionately. The natural selectivity of the environment, in this case represented by lava rocks, would have chosen the survivors on the basis of color. A new kind of heron would emerge.

As the Galapagos Islands age, as more and more of the

rocks are reduced to soil, and as the soil gradually permits the development of more green plants on the barren areas, it may no longer be advantageous for a heron to be black. The environment may then be selecting not dark birds but green ones – "throw-backs" to the original form which will return through the generations. More of these will survive than are presently permitted, and a majority of the population will again be more colorful. The dark birds will return to the minority, and perhaps for a while disappear altogether.

The significance of the Galapagos archipelago does not end with the fact that its animals have become different from their opposite numbers (and antecedents) on the mainland. Things have gone a great deal further than that. A new mutation such as color, which may be beneficial to an animal on the Galapagos, would probably have been suppressed on the mainland. On some neighboring island a mutation may have gone in a different direction. Given sufficient time, and the fact that animals such as small lizards are not successful swimmers, the populations on two or more islands may begin to differ from each other, as well as from their mainland ancestors. They become different animals. Eventually, they become so different that even if they were united, they would no longer be capable of interbreeding. That is our concept of a species. The operative phrase is "reproductive isolation."

Species are plastic and consist of many varieties. The song sparrow of North America includes more than thirty different geographic varieties (races, or sub-species) of the bird. Each is potentially a new species. The point strikes home when you arrange on a table a number of skins of song sparrows from a museum collection. You might arrange them by color – from the palest to the darkest – or by size, or by bill characteristics. But whichever way you arrange them, you are struck by the fact that no two of them are *exactly* the same. If your life depended upon your ability to select one and say, "This is the song sparrow," you could not do it. You could only say, "This is *a* song sparrow."

Regardless of their variation, they all maintain their integrity as song sparrows. Each is a genetic experiment, and if one of them should be isolated long enough it would undoubtedly evolve into a distinctly new bird – no longer a song sparrow, but something else. In the Galapagos there are now four kinds of mockingbirds where once there was one. No doubt one day there will be new kinds of song sparrows. So "species" is not, in fact, a reality. It is merely a convenient description of a point in time. The true reality is in the individual animal, with his genetic inheritance, claims the famous zoologist Ernst Mayr; and no individual can be identical to any other. It is the latest in the unending series of chance genetic experiments.

There are in Canada two kinds of flickers – rather large, brown woodpeckers which do a good deal of foraging on the ground. The eastern form has yellow feathers in its wings and tail; it is called the yellow-shafted flicker. The western bird is red-shafted. The eastern form lives from the Maritimes to the Yukon, but not in British Columbia. There it is replaced by the red-shafted bird.

Most authorities recognize these two as distinct and different species. Yet they commonly "hybridize" where their ranges overlap in Alberta. There, resulting offspring may be either yellow-shafted or red-shafted, or a mixture of the two. It would seem that the two flickers have not yet diverged sufficiently; and were they to unite, their "specific" distinctions might quickly disappear.

There are in the world today about 8,600 distinct species of birds. Of mammals there are about 5,000, of reptiles about 6,000, of amphibians some 1,500, and of fishes something in excess of 20,000. Thus the existing number of vertebrate species is something over 40,000. Of which man is one. Forty thousand green searching tendril-tips in the vertebrates alone. Numbers of species of invertebrates are astronomical; there may be more than a million different kinds of insects, to say nothing of their innumerable relatives. We humans represent only one experiment among millions.

No species is "more advanced" than any of its predecessors. We live in an arbitrary moment on a long continuum, and species existing today are those which have in their genetic make-up the attributes for survival in the various world environments as they chance to exist at this moment. For an animal to have been extant at any given instant over the ages means that it had in its genetic make-up at that time the necessary attributes for survival, and its environment allowed it to survive.

"Survival of the fittest" is the survival of those animals best equipped – best adapted – in the context of their environment at a given time and in a given place. But since no environment is static, individuals, species, groups of species and whole faunas have found themselves maladapted to new situations and have given way to other animals which, quite by chance, found themselves the possessors of a genetic inheritance which was suitable to the new conditions which arose.

The new lava-colored "green" heron of the Galapagos represents a classic type of adaptation. The rocky environment favored natural selective screening for color. Adaptation can take many forms; it can involve the physiology and even the behavior of an animal in addition to its anatomy and morphology. Because we are such close mammalian relatives, the giraffe has the same number of vertebrae in its neck as we have, but it has taken a different form. The African bush environment has allowed long-necked and long-tongued giraffes to make a very successful living from browsing on acacia trees.

One of the penguin's most obvious characteristics is its inability to fly. If circumstances should change, the bird which has "abrogated" the power of flight could be seriously endangered. But in the antarctic, penguins have no land-based predators, and they are fairly safe in the water. A few fall prey to leopard seals, but not enough to make any difference to the population as a whole.

It is not entirely true, however, to say that a penguin cannot fly. It can no longer fly in the *air,* but it does fly in the water. Penguin wings have evolved into stiff, sturdy flippers, and penguins are as graceful underwater as fishes or seals. They are descended from flying birds, but their way of life no longer depends on aerial prowess. They have relinquished air flight for mastery of the water. In the same sense a seal or a sea lion has given up the mobility on land which once was the property of its doglike or bearlike ancestor in exchange for a suppleness and elegance in the water which is unmatched in the animal kingdom, save by the penguins.

These changes were not "purposeful." The seal and the penguin did not "decide" to re-invade the water which gave rise to their earliest ancestors. They clearly liked the water, as the ducks and grebes and otters do today. Chance mutations which made them even more adept at handling themselves in the water were allowed to persist because they increased the animals' efficiency. (A sea otter is better adapted, and thus more committed, to the water than a river otter or a mink.) As they became more and more specialized for swimming, they became less agile on land.

The world-renowned vertebrate paleontologist and teacher, Alfred S. Romer, explains the sequence of events which changed the ancient crossopterygian (lobe-finned) fishes into the world's first land animals, the amphibians. The Devonian period, the Age of Fishes, is thought to have seen marked changes in world climate, and thus in local environments. At some time, widespread drought is thought to have gradually dried up great numbers of ponds, lakes, and marshes. This was hard on the fishes. Some of them, however, could breathe air, and the lobe-fins also had thick, fleshy fins almost like little limbs.

As these strange, stumpy-finned fishes wriggled, struggled and crawled over the mud, they were making their way overland in search of water. Gradually their swollen, sturdy fins became stronger and eventually evolved into legs. However, they had no notion of remaining on the land, much less of becoming amphibians.

Charles F. Hockett and Robert Ascher in their famous article "The Human Revolution" have expressed this conservatism of adaptation – the force to go on leading the old ways of life under changed conditions – as "Romer's Rule": "The initial survival value of a favorable innovation is conservative, in that it renders possible the maintenance of a traditional way of life in the face of changed circumstances." Eventually the innovation, if successful, becomes one of the means of leading a new way of life in a new kind of situation, in the crossopterygians' case, on the land. Evolutionary conservatism also has implications in the story of human evolution.

A species or a group of species frequently becomes so narrowly adapted to a set of conditions or a way of life that it becomes *over*specialized. A koala bear, for example, eats eucalyptus leaves and nothing else. One conspicuous result of this is the absence of koalas from most of the world's zoos, which cannot supply the fresh eucalyptus leaves. Another possible result would be the total disappearance of koalas, should some blight wipe out eucalyptus trees. The ivory-billed woodpecker, largest of its tribe in North America, was a victim of such overspecialization. It depended on insect food in the oldest bald cypresses of southern swamps. When the swamps were drained and the trees cut, there were no more ivory-billed woodpeckers.

One of the mysteries is the disappearance of the dinosaurs, once the dominant creatures of Earth's land masses. What caused these gigantic beings to vanish almost without any survivors save the crocodilians and that lonely New Zealand waif, the tuatara? Being reptiles, dinosaurs could tolerate ambient temperature variation only within narrow limits. One theory is that the heat became too intense for them: the internal testes of the males heated to the point of sterilization. (The mammalian scrotum is a sperm-cooling device.) Or the world got too cold for them. No doubt the last carnivorous dinosaurs disappeared with their vegetarian fellows, which represented their food supply. The vegetarian dinosaurs may not have been able to endure widespread flooding, or drought. The dinosaurs were survived only by some of the lesser reptiles and by the mammals and birds, which had long since become established.

Great size is itself an overspecialization. It demands immense space and food supply. Another conspicuous over-specialization might be the ludicrously long tail of a male peacock, which must make him extremely vulnerable to predators. Or the apparently unnecessarily wide antlers of a moose. Or the astonishingly large brain of a man. Beyond a certain point, moderation is not only a virtue but an evolutionary necessity.

The most common question about the dinosaurs is "Why?" – not only why they became extinct, but why they got to be so large, and why they evolved in the first place. In nature, asking "Why?" is fruitless. What matters is "How." At any given moment in the long geological continuum, the individuals and species and faunas on Earth are those which because of chance events of climate, geography, and genetics, happened to be the most suited to the particular conditions of that moment. In their time, they *were* the fittest. But times change.

The concept of purpose and direction in evolution as expressed in an anthropocentric pyramid or tree of life is untenable. The world of living things is not a pyramid or a tree or a ladder, with a hierarchy from "primitive" to "highest." Like all worlds, the world of life is a sphere. Each point on its surface is equidistant from its center. Just as Washington and Hanoi are equidistant from the center of Earth, the bald eagle, the Vietnam pot-bellied pig, and the human species, are all equidistant from the center of the sphere of life.

We can reduce the sphere to the two dimensions of a flat lichen community. From its nucleus it expands forever outward toward every point of the compass, just as the cosmos expands into infinity. The concept of direction as expressed in a man-dominated pyramid or tree of life has no relation to reality. To regard the present condition – a monoculture of one dominant species of large mammal – as the inevitable result of a "master plan" is to reveal ignorance and insensitivity. Like other hyper-specializations, the arrogant human brain will eventually be dealt with.

The grandeur of life and life's processes lies in the exquisite opportunism of living things and in the intractable and dispassionate influence of changing environments.

It is sufficient "purpose" for a deadly nightshade vine that its violet and yellow blooms please me in the summer and that its ruby fruits feed some wandering pine grosbeak in the winter.

Beyond that, let us not look for purpose in the cosmos or in life on Earth. It is enough to see, to value – and to savor our capacity for wonder.

CHAPTER SIX

Measuring the Upright Man

Through the tortuous courses of evolution, the key factor has been adaptability. The living thing that cannot become adapted to changing circumstances in its environment is screened out. The animal that becomes overspecialized in a certain set of conditions drifts into extinction if those conditions change. A species can be wiped off the face of the planet, or it can evolve into something new, something better equipped to survive the slow but inexorable process of environmental change. Of all the diverse assembly of vertebrates, and the orders of mammals that have overcome varying environmental vicissitudes, among the most successful have been the primates.

A primate is more than a bishop. No doubt it was inevitable that we would ascribe the first order of rank to the mammalian order to which we ourselves belong. How could it be otherwise? In addition to bishops and lesser persons, however, the order of living primates includes such intriguing blood-relations as the tree "shrews" of the Far East, the lemurs, woolly lemurs and aye-aye of Madagascar, the slow loris, pottos and galagos of Africa and the East, the tarsiers of the southwest Pacific, the monkeys and marmosets of Central and South America, and the monkeys, gorillas, gibbons, orang-utan and chimpanzees of Africa and Asia.

Primates appear to have evolved shortly after the midway mark of mammalian history, about 70 million years ago. Scientists have placed primates in a position on the phylogenetic scale immediately following the insectivores and bats and preceding the toothless anteaters, sloths and armadillos, and the pangolins. (Their objectivity in placing the primates there rather than at the end, which is usually regarded as the most "highly" or at least most recently evolved position, is noteworthy, and illustrates why family "trees," with the topmost growth bud inevitably representing man, can be gravely misleading.)

The number and variety of our living primate cousins illuminate our own origins. Over 100 years ago, T. H. Huxley noted that the primates as a group appear to represent a series of more and more "advanced" forms, and that we can infer from them a reasonably accurate picture of the dim evolutionary past. Existing primates enable anatomists to draw some inferences from the increasing number of fossil remains. However, none of the living primates is thought to represent a specific stage in the evolutionary development of the group as a whole.

Of all the characteristics which primates have in common, none is more important than the hand, with opposable thumb. The tree "shrew" does not have this facility, and some believe it should therefore be assigned to the insectivores.

A tree "shrew" has the appearance of a slender little squirrel with a long feathery tail, enormous eyes (it is nocturnal), and a thin pointed nose. The fact that experts are undecided whether this tree-dweller of Malaysia is a primate or an insectivore is illuminating, for it indicates the blurred transitions through which all organisms evolve.

Small mammals were in existence almost since the days of the earliest dinosaurs, but it took time for them to emerge from the dark and tangled security of dense tropical forests. Through the ages of geological time, they became remarkably skillful at surviving, and proliferated into many forms. About 70 million years ago, mammals had diversified to the point that there were in evidence the progenitors of the primates. Judging by existing members of the group, these were probably night-going creatures resembling squirrels and tree "shrews," agile, and expert tree-climbers. They must have eaten whatever vegetable and animal food they could find. Without opposable thumbs, they probably manipulated food with their forelimbs, in much the same way as a chipmunk does today.

These creatures must have required exceptional eyesight in order to judge the distance of leaps from branch to branch. Natural selection for depth perception would undoubtedly be rigorous: it would screen out the small furry creature whose jumps were inaccurate, and favor more rapid movement of the eyes away from the sides of the head, as in a squirrel, and toward the front of the head, as in a monkey.

As the arboreal life became more specialized in its adaptations, some of the primates developed an even wider range of binocular vision and depth perception. Not only were they able to jump with accuracy, they were also able to fiddle with things in their "hands" and see what was going on with both eyes at once. No doubt the fingers and thumb must have begun to emerge at this stage. The hands of modern lemurs give us a clue to this development. Also, improved vision led to advancement in parts of the brain and in its size.

The sense of smell does not appear to have been an important factor to the larger primates. Smaller members of the tribe, such as lemurs and galagos, do rely on it in territorial scent-marking; but the olfactory apparatus has become less and less relevant to the life styles of monkeys, apes, and man. We may *think* that we react with selective sensibility to the smells of flowers, or food, or sex, but these are the coarsest of responses. A dog's world of scent is a dimension which we have totally forgotten.

For millions of years, jungles were full of the earliest primates. But it was long before anything resembling a monkey appeared. By that time, Africa and South America had begun their slow drift away from each other. This may have left some of the earliest proto-primates on each of the drifting chunks. The South American monkeys are vastly different from their Old World counterparts. There is no doubt that they had ancestors in common, but those ancestors must have existed as long ago as 150 to 100 million years. So far as we know, the New World produced no ape-like creatures. The American primates must therefore be exonerated from any hint of implication in the evolution of man.

One of the major criteria for the identification of animals included in the group of newer primates – monkeys, apes, and man – is the reduction in the number of teeth to 32. The oldest forms which thus qualify are dated at approximately 30 million years ago. This seems to have been the time when monkeys and apes began to go their separate ways (man is much more closely related to the great apes than to the monkeys). At the same time, even the ape line was beginning to diverge from the mainstream, and the ancestors of the gibbons were beginning to look like real gibbons. Gibbons, the most acrobatically aerial of the apes, have developed extraordinarily long arms and fingers for swinging from branch to branch. They cannot walk very well, but they run well, with their elongated arms either waving in the air for balance or collapsed and folded turban-like over the head.

Twenty million years ago, before there were apes or monkeys in recognizable form, there were some antecedents. Of these the most important was *Proconsul,* now thought to be an ancestor of the modern gorilla and chimpanzees, but not of ours. Most likely to be directly in our own line is *Ramapithecus,* which has been found in both India and Africa. He is thought to have walked upright.

Uprightness is not necessarily the measure of the man. All kinds of animals can stand on their hind legs, and some are even able to move quickly in that stance. Ostriches and penguins are examples, and in the world of mammals, the bears and a number of primates. Apart from the gibbons, the great apes are not very good standers and walkers, although they are able to when the occasion demands, or when they are taught. The most arboreal of the great apes is the Asian orang-utan, which rarely ventures onto the ground. One young orang in a zoo, trained to stand up on its hind legs, moved slowly and awkwardly with a strange stiff gait, its limbs swinging straight from the hips like those of a peg-legged man. It was christened the "Lock-Knees Monster" by my friend Jim Murray. The poor creature was really not comfortable, but in the gentle good nature so characteristic of orangs, it humored its audience. Chimps and gorillas can walk a little better, but it appears to be tiring for them.

In his famous paper "Food Transport and the Origin of

Hominid Bipedalism" Gordon W. Hewes commented that we "should not overlook the possibility that all cultural systems have a certain vested interest in making the distinctions between man and other animals as sharp as possible. Humanity is self-consciously bipedal." Wilfred Neill, the crocodile authority, states: "Perhaps to man's prehuman ancestors, the first advantage of an upright posture was its effect upon fearsome mammalian predators of that day."

As those who have encountered one know, it is possible to repel, at least momentarily, an angry bull sea lion by maintaining an upright position, which makes you considerably taller than he is. In certain circumstances there is among animals an important aspect about relative size as manifest in *verticality*. The bipedal primate was already well on the way to "sensing" a substantial difference between himself and the rest of the animal kingdom.

How did bipedalism arise? Most primates are creatures of the forests, and beyond the necessity of skipping a few paces along a limb on hind feet while looking for a new hand-hold, the need for walking on two feet would rarely arise. There follows the question: "Why come down out of the trees in the first place?"

Perhaps sheer size and bulk drove our ancestors onto the ground, as it has the largest of contemporary male gorillas who, because of their weight, rarely do much climbing any more. But our ancestors were not comparable in size to the gorillas. Later, there were large species of man-apes, but the primates who forsook the trees for the plains of Africa were believed to be much smaller than we are. Weather is considered to have been the determinant. Environment changed, and various kinds of primates became adapted to the new circumstances.

Sweeping climatic change is said to have broken up the vast stands of tropical African jungle into more or less discrete patches of forest, like islands in an expanding sea of grass. Many of the larger primates remained in their patches of forest, and no doubt in due course they gave rise to today's great apes. Others, however, found it necessary to move out of their shrinking groves and to venture onto the open prairie in search of another suitable woodland. (This is comparable to the lobe-finned fish of the Devonian, in his hour of extremity, struggling to make his way to a better pond.) Many of the primates found new forests in which to pursue the old ways of life; others were incapable of reaching the right kind of woodlands, and remained stranded on the ground. They failed to handle the habitat change in the old, conservative way, and they were forced to handle it another way – through adaptation.

We can picture a form emerging, with slightly shorter forelimbs and longer legs than a modern ape – perhaps something like a baboon. It moved around on the ground on four legs, occasionally standing up to peer over the grasses for enemies. When it recognized a threat, perhaps it scampered away on its hind legs, managing to keep its head above grass level in order to keep track of the danger.

Bipedalism may have arisen in this way, but the feeling today is that it evolved in response to the need to carry food from place to place. Running on the hind legs released the forelimbs for carrying. (There are those who say that this is abundantly true, but what the little monsters were carrying was not food, but weapons.)

Gordon Hewes believes that the only possible reason for a four-legged primate living on the ground to become a habitual two-footed walker would have been "food transport from the places where food was obtained to a home base where it was consumed." Primate hands are better for carrying than primate jaws are. Jaw-carrying is for cats and dogs, and dog-like primates such as baboons. Even baboons will scuttle along on three legs while holding some tid-bit tucked under the arm like a football. Small forest monkeys such as vervets will often run short distances with booty clasped to their chests.

There is no doubt that our ancestor was substantially a meat-eater. The average primate does not have the right kind of jaws and teeth for cracking bones and slicing tendons, and neither do we. But our ancestor might well have stolen a chunk of meat from the kill of some large carnivore and made off with it to a safe place where it could be picked at leisurely. There must have been many such kills. In those days, the plains of Africa supported an even wider variety of large vegetarian animals than they do today; and there were many predators, such as big cats, dogs, and hyenas.

The little primate, becoming more and more upright, could probably frighten away the jackals just by waving at them. Of course it must have been a different case with the hyenas, which then as now were big, dangerous animals. One can imagine the scavenging primate, in a fit of excitement, shaking a bone, a stick, or even rolling a stone or flinging a piece of turf at the intruding hyena. Slowly our forefather was emerging.

Primates which have become increasingly numerous in museum collections are the australopithecines, and these are closely tied to the genealogical history of man.

Australopithecus existed in many sizes and had equally varied life styles, judging from the fossils. The famous *Zinjanthropus* (now called by its discoverers *Australopithecus*) *boisei*, found at Olduvai Gorge in Tanzania by Louis and Mary Leakey, has been dated at 1.75 million years. What is believed to be the same species was unearthed at Lake Rudolph in Kenya by the Leakeys' son, Richard, in 1970. This one has been dated at 2.6 million years. These were large for australopithecines – they may have been five feet tall. They had immense jaws and grinding teeth twice the size of ours, but their brains were not much bigger than those of gorillas. Their teeth would indicate that, unlike some of the other australopithecines, they were strict vegetarians.

F. Clark Howell describes *Australopithecus* as "a slender four-footer, weighing under 100 pounds." He appears to have been nimble and erect, and, like modern man, he probably ate most everything that came his way, both meat and vegetable. And he had tools. They were mostly pebble "choppers" which had been deliberately flaked. These have been found in great numbers at Olduvai Gorge, together with bones that had been broken to expose the marrow. These tools are relatively "recent" (a couple of million years old). But the use of tools, as opposed to their making, goes back much farther.

The most widely publicized non-primate tool-user is a finch of the Galapagos. In the Galapagos Islands, there are no woodpeckers, tree creepers, nuthatches, or other birds adapted to foraging in the bark of trees. But there is plenty of loose bark in the groves of trees which grow on some of the islands. The bark is filled with the larvae and pupae of insects. In the absence of other birds to do the job, a finch has taken over.

No finch has the special beak of a woodpecker, creeper, or nuthatch, nor the long barbed tongue characteristic of most woodpeckers. So the Galapagos bird must do its foraging some other way. The tool-using finch pecks away at the bark with its bill (which is often open, giving the appearance of striking with the lower mandible) until it finds a food item. Then, unable to reach it with bill or tongue, it flies to a nearby cactus plant and breaks off a spine – or perhaps a twig from some other tree – returns to the site of its activity and pokes around with its small "toothpick" until the prey item is dislodged. Then it drops its probe and snaps up the morsel. Most remarkable of all, perhaps, is the fact that the bird will "fashion" a tool to its liking by breaking a longer twig down to more manageable proportions. Recently the crow of New Caledonia has been observed to probe with sticks in similar fashion.

Another tool-using bird came to light through the observations of Hugo and Jane van Lawick-Goodall. They discovered that the small white Egyptian vulture of East Africa will use stones with which to crack the tough shells of ostrich eggs. (Through experimentation, they also found that the vultures will pick up the eggs of lesser fowl and hurl them to the ground to break them.) One might speculate that gulls, which often drop mussels on rocks to crack them open, might one day decide to drop rocks on mussels.

Tool-using of a different kind is practiced by the handsome satin bower bird of eastern Australia, which uses a tool in the form of a "paint dauber" – a wad of leaves with which it paints the interior of its thatched bower. The color is a blue pigment made of charcoal.

Only one non-primate mammal is known to use tools. The sea otter of the Pacific coast of North America uses small flat stones to crack shellfish which it brings up from the sea floor. Lolling at the surface on its back, the sea otter places a stone on its chest, then knocks the shells against it until they break open.

The role of tools in the early days of proto-man has perplexed anthropologists. Which came first – the tool, the upright stance, or the burgeoning brain? Which gave rise to which? There are conflicting theories, but they probably developed simultaneously. Environments evolve and animals evolve along with them, with different adaptations developing simultaneously. Probably stance, tools and brain all developed at the same time, constantly feeding back to each other as though by systems "loops," allowing the whole complex being which was to become human to develop at the same relative speed as all of its parts.

Tool-using is different from throwing, which many primates seem to do habitually. One can be showered with sticks and other debris, including feces, released by spider monkeys in the tropical American jungle. An angry chimpanzee in captivity will bowl chunks of turf at its tormentor.

A. Kortlandt, the animal psychologist, has noted remarkable skill and accuracy in throwing by zoo chimpanzees, to the extent that they were able to hit a moving target by aiming in front of it, thus contriving to smash car windows. He has also commented on a young zoo chimpanzee which "attained a hitting accuracy considerably better than that of average children of comparable physical development." Chimpanzees born and raised in captivity have been known to throw "anything at hand" at an aggressive-looking leopard or tiger. It would seem that throwing is deeply ingrained in us primates, if not for purposes of obtaining prey, at least for driving off enemies. A recent film of wild chimpanzees showed the apes taking large sticks with which to attack a dummy leopard suddenly revealed to them.

For maximum throwing skill, however, it helps to be strongly bipedal. Standing on two legs makes it much easier to throw accurately and with force. Any quarterback knows how difficult it is to throw from a sitting position. Only the human primate has produced such throwers as Joe Namath and Dizzy Dean. There is a degree of coordination and balance involved here which is quite impossible for a nonhuman primate. By comparison with our relatives, some of our physical skills have been vastly underestimated.

The view of Hewes was that true bipedalism arose in this primate as the result of the practice of scavenging meat from the kills of large carnivores, and carrying these bits and pieces away to some home base for subsequent consumption. Tools came later. The opposing school, exemplified in the work of Sherwood L. Washburn and others, holds to the opinion that it was the use of tools which led to the two-legged stance, not the other way round. This is based, quite logically, on the fact that the apes are known to use tools, and that apes are by no means habitually bipedal.

Apes certainly do throw sticks and stones, however, and as Howell puts it, chance success that they may have enjoyed by doing this "may have led to a dawning realization that rocks and clubs were useful as weapons, even for bringing down small game." From there, the theory goes, it was merely a matter of development toward greater efficiency with sticks and stones. This would include carrying the missiles with one, and beginning to run on two legs with them as we know apes and monkeys are capable of doing. The small terrestrial primate was better at it than the monkeys and apes were, and gradually the habitually erect stance evolved – due to the carrying of crude weapons.

Another issue for the paleoanthropologist is whether the stick or rock was in fact a weapon or merely an "implement."

Ashley Montagu, the noted anthropologist, maintains that all weapons are tools, but not all tools are weapons. No tool used for killing prey, he feels, is a weapon; the only weapon is the instrument that is used for attacking *one's own species*. A spear, Montagu says, is an implement when it is used to bring down an animal, but it is a weapon when used to strike a man. We must then determine what these early australopithecines were doing. Were they killing other animals, or were they killing each other?

Here the specter of cannibalism looms – the devouring of one's own species. Cannibalism is *not* the killing and eating of another species, even a closely related one. This is where the "man and nature" syndrome manifests itself: man in one box and "nature" – which means all non-human species – in another box. The gentle soul who sees a hawk eating a fowl calls the predator a "cannibal" because it is a bird eating another bird, neglecting to remember that the hawk and the fowl are as different as the gentle soul and the lamb that provided last night's dinner. Cannibalism is rare among birds and mammals, though it is relatively common in cold-blooded animals. It does occur in hawks and owls, where it is in fact fratricide. It occurs occasionally among hyenas and bears, but it is everywhere the exception.

There is little evidence to show that the australopithecines were cannibalistic; indications of it may or may not emerge at some time. But having in mind the general rule, one may be confident that it was rare. The point is, however, that for an *Australopithecus* to make a meal of a *Paranthropus* is no more cannibalistic than it is for a leopard to carry off and eat a Siamese cat, or for a chimp to catch, kill and eat a colobus monkey. They are different species: it is a simple predator-prey relationship. It is probable that there were several australopithecine species, and it is no less probable that from time to time they killed and ate one another. But all the present evidence points to the earlier forms of *Australopithecus* as concentrating on very minor game.

Food gathering and scavenging is a hit-and-miss existence, with the forager settling for what he can find. In total, however, it probably amounts to a well-balanced primate diet. Small lizards, insects and their larvae, birds' eggs and nestlings, the occasional small mammal and whatever chunks of meat could be stolen from carcasses, would nicely supplement the primate mainstay of berries and fruits, nuts, green shoots and tubers. A generalized diet of this kind requires generalized teeth – not the shearing and slashing teeth of dogs and cats, nor the grinding and masticating teeth of ruminants.

Apes (except for gibbons) have notably large, sharp canine teeth, especially in the males. These serve a twofold purpose – for tearing away at the tough covering of much of the vegetable matter on which they depend, and for combat, which includes defense. So far as evolving man is concerned, it is believed that once weapons were developed to a point at which they were more effective than teeth, then natural selection took over and favored smaller and smaller canines.

There is evidence, however, that the canine began to shrink *before* the australopithecines became habitual tool-users. Even *Ramapithecus*, 14 million years ago, shows signs that the teeth were changing. If, at that stage, teeth were becoming less important, it would indicate that something was replacing the need for them. Perhaps primitive weapons were originating, with cutting implements to reduce meat to bite-size chunks. This seems to have been the case with the australopithecines, and one wonders whether it might have arisen a great deal earlier. If so, the separation of man and nature has far deeper roots than we may have thought.

Both the teeth and the skull of *Australopithecus* had remarkably human characteristics. His upright stance meant that his skull was balanced at the top of his spinal column. He had short, broad hip-bones and substantial buttocks to hold him upright.

The earliest australopithecines very likely ran better than they walked, but their leg bones gradually lengthened, their hips widened, and they became able to slacken the pace somewhat. At no time would they have been able to perform the slow march of the Grenadier Guards without falling over, but they improved, and were able to walk without losing too much of their stability and agility. (It is noteworthy that despite our modern structure, many forest-dwelling peoples today climb trees extraordinarily well.)

It would appear that australopithecines of one kind and another persisted well into the era of men. Where the branching-off took place has yet to be determined, but the earliest of the australopithecines gave rise to both the last of their kind and the first of true men. Whether the earliest *Homo* actually exterminated his blood-relations is not known, but it appears to be highly possible, having in mind that there is no room in nature for the coexistence of competitors. Their ways of life must have been strikingly similar. The two forms, *Homo* and *Australopithecus,* coexisted at least a million years ago in Africa. If one of Richard Leakey's latest finds at Lake Rudolph in Kenya proves to be a prototype of *Homo,* they coexisted over two million years ago.

In 1960, at Olduvai Gorge in Tanzania, the senior Leakeys unearthed a fossil to which they gave the name *Homo habilis.* It was dated at 1.75 million years. Much dispute surrounds this discovery, but certainly it is a far different primate from the australopithecine remains at the same site. Some describe the "handyman" as a transitional stage between the australopithecines and the first known man on which there is unanimity, *Homo erectus.* The latter is the same species we used to know variously as *Pithecanthropus erectus* or *Sinanthropus pekinensis.* The "Java man" and the "Peking man" of early texts have turned out to be one and the same, and have been posthumously promoted into the headiest of company, the genus *Homo.* All this underscores the fact that true man has been on the planet much longer than we used to think – in fact, he existed concurrently with the last of the man-apes (or ape-men).

The oldest known example of *Homo erectus* is our ancient acquaintance from Java, who lived more than 700,000 years ago. There is little question that he originated in Africa. His remains have also been discovered from Algeria to South Africa, and from Germany to northern China. He endured for a very long time – so long, that his career overlapped not only that of the later australopithecines but also that of modern man. That he was a transitional form between the two is no longer seriously in doubt; it is now a matter of fitting in the pieces.

Homo erectus was just that – erect man. He stood bolt upright and he was an excellent stroller – so good that even the experts, presented with his leg bones, cannot tell whether they belong to *erectus* or to *sapiens.* The accumulated fossils show that he was about five feet tall, and robustly built. He had a brain about twice the volume of that of the average australopithecine. *Homo erectus'* brain averaged a little less than 1,000 cubic centimeters; but the largest discovered so far reached 1,300 c.c., well into the range occupied by present man.

Erect man used his expanding brain in ways not employed by australopithecines. He explored tool-using further; and before he faded away, his brain had led him to hand-axes, cleavers of fine proportions, and wooden spears. He also recognized the values of fire and advanced that knowledge. He was able to communicate with his fellow man, and had a faint stirring of a cultural tradition.

In view of the growing collections of fossil remains and the apparent overlapping shapes and sizes of skulls, it becomes impossible to assign a precise date to the final disappearance of *Homo erectus.* The last typical specimen is more than 300,-000 years old, somewhat coincident with the appearance of men of our own species.

The difference between ourselves and the preceding member of our genus is not in kind; it is in degree. We have smaller features, more bulbous foreheads and less bulging brow-ridges; we are more slender, perhaps somewhat taller. On average, we have larger brains, encased in rounder skulls. We are probably not as strong as he was, but then we do not need to be. All in all, the external difference is not conspicuous.

In the framework of geological or evolutionary time scales, the modern human brain grew to its present dimensions and complexity almost overnight. It evolved far more quickly than the development of our physique, which occurred over millions of years. The brain of a physically highly evolved australopithecine one million years ago was little more than one-third the volume of our own. The brain of an average *Homo erectus* only half a million years ago was on average about two-thirds the volume of our own. Put another way, the brain of the australopithecine virtually doubled in size while it was evolving into the brain of *Homo*. The increase from *erectus* to *sapiens* is smaller. Why this incredible, explosive change from the last man-ape to the first man? It must be attributable to social behavior. The measure of a primate, after all, even a human primate, is not in what he looks like or what he is or what he says he is. It is in what he *does*.

CHAPTER SEVEN

On Preying Together

Most primates are social creatures, and there is little doubt that our pre-human ancestors lived and travelled in bands, as monkeys and apes do today. There is wide variation in the social systems of different kinds of primates, and this seems to be determined by the environmental circumstances within which the animals find themselves rather than by the genealogical relationships of the various species.

One of the more conspicuous behavioral traits of our cousins the gibbons is their noisiness. They whoop up a resounding storm in their dense habitat of tropical Asian jungle. Also noted for their vocal strength are the howler monkeys; big males sound, at any distance, almost like lions or jaguars. But the two are only very distant relations: the gibbon is an Old World ape and the howler is a New World monkey. They are both primates, but here their relationship ends. Both must depend upon their vocal power because they live in thick equatorial forests where it is almost as difficult for sound to penetrate as it is to travel through the vegetation. Keeping in touch with one's immediate group or with other groups is much easier to do vocally than any other way, given the constraints of the environment.

When we were still making a life in the trees, sound communication may not have been as important as it is to the howlers and gibbons. We probably lived at the edge of the woods, in an environment similar to the one baboons inhabit today. But when we became terrestrial, chattering may have helped keep our groups together. (Small birds such as chickadees, warblers, and kinglets, which generally travel in flocks, keep up a perpetual chipping and peeping as though to ensure that everyone knows where everyone else is at a given time, and individuals do not become separated from the main group.)

It would appear that primates group together primarily to protect the younger members of the band. It matters little whether this is conscious or not, for that is the effect of it. One would think that the most important role of sound communication in the earliest days was as warning. A loud alarm note would alert everyone within earshot of danger. Baboons are generally silent, but when one of the group detects a leopard or other interloper in the vicintiy, the ensuing "bark" has an uncannily human quality. Such a device must have been employed for millions of years in view of the weakness of primate scent and the sharpness of primate eyesight.

The little primates, proto-men, probably nested in trees at night, as chimpanzees, gorillas, and orang-utans still do. Orangs very rarely come down to the ground, because they are awkward there and the forest floor offers them little inducement for exploration. But chimpanzees spend a great deal of their time there, and the gorillas almost all of it. Yet at sunset they return to the trees, where they build individual sleeping nests wherever they find themselves.

Night-nesting is clearly a response to the ubiquitous presence of a variety of ground-based predators which, with the exception of the leopard, are not good climbers. Even the leopard, though he is reasonably agile in the trees, is not likely to go hunting there. It follows that the heat is off the large apes for a substantial part of every day – half of it, in fact. (Large male gorillas, we are told, do not bother to ascend into the trees. Or rather, they have grown too big to navigate them. They are fairly secure in the nests which they build on the ground. It is hard to imagine a predator in the African jungle with the temerity to take on a male gorilla.)

Observers such as George Schaller have noted that the great apes (in this case, gorillas) climb into their sleeping nests well before darkness falls, and do not leave them until after daybreak. They live in equatorial regions, where day and night are roughly the same length. They sleep for a full twelve hours every day – which would seem to be more rest than they actually need considering that a gorilla's normal daytime routine is anything but active.

One wonders at what point our forebears reduced the

twelve hours' sleep to one-third of the 24-hour cycle. Perhaps the development of a high degree of confidence in our survival potential brought about this departure from tradition. Four hours awake in the dark would be unthinkable to a chimpanzee or a gorilla. Perhaps it came with the advent of controllable fire. However it arose, that extra period of daily activity must have had incalculable social consequences. Four extra hours for eating, for tool-making and maintenance, for sex, or for communication. Four extra hours for the conglomeration and cementing of tradition – for learning, for language, and, at some point, for ideas.

It is difficult to conceive of proto-man being sufficiently daring to witness nightfall in the absence of fire. Yet other forms of defense – such as effective weapons or well-developed systems of group action – may have allayed the age-old fears. Whatever the mechanism, the abandonment of the twelve-hour sleeping cycle meant at least some feeling of immunity to natural enemies.

Of all the natural enemies of the primates, the cats are at the head of the list. Lesser cats take their toll of lesser primates, but among the baboons and apes there is no threat like that of the leopard. Tigers occasionally take langurs (long-tailed forest monkeys), but lions rarely if ever stoop to such paltry fare. In addition to the cats, there are the giant snakes, the constricting pythons and others, which take a certain number of primates. Also, there are birds, such as the great monkey-eating eagle of the Philippines. And finally, the primates themselves.

On the basis of present-day primates, fear of darkness seems to be well founded. With a few exceptions – such as the South American night monkey – apes and monkeys do not function after dark. Night is a bad time for a primate, especially a baby one, and it seems obvious why the little primate of our own species demands a night-light in the bedroom, and fears the terrible unknowns of the darkness. One's helplessness is never so conspicuous as it is in darkness.

A substantial case could be made for the existence of a

deeply ingrained primate tradition about darkness – and about cats, and snakes, and other threats one cannot see. Our noses long since ceased to be of any defensive value. At night we are left with only our hearing (and our quickening imagination). It is easy to picture, millions of years ago, moonless nights of terror. There have been African nights when I have remembered them.

No ancient fear is more deeply etched in our primate memory than the fear of snakes. A snake does not make himself visible until that last split-second when we are about to walk on him, and then we congeal with terror. It is a long, long memory which goes back to the time we were small animals and snakes were more common than they are today. Usually, by the time you see a snake, the initiative is his. We dislike having our initiative removed, and we especially resent and fear things hidden from our view. The claptrap which attempts to relate women's fear of snakes to their (highly questionable) fear of the male penis is utterly unfounded. Women, who are ground-dwelling primates, have perfectly good reason to be afraid of snakes. So do men.

One would imagine that it was a long time before the proto-man-apes foraged at night. The new primate was a diurnal animal, operating in broad daylight, chiefly by means of his excellent eyesight. From his tree-dwelling ancestors he had inherited the intestinal flora to handle not only vegetable material but also animal protein in various forms. In the early stages, this animal material was probably limited to birds' eggs and nestlings, insect larvae and other minor fare. Later, primates such as the early australopithecines apparently took to scrounging, scavenging and stealing morsels from the kills of large carnivores such as lions and hyenas, and perhaps wild dogs. Lions or their equivalents of the time probably contributed more to the scavengers than did hyenas or dogs; the latter usually bolt their meals so quickly and completely that there is little left for anyone else.

However, a defenseless four-foot primate would be unlikely to walk up to a pride of dozing lions and attempt to make off with a choice cut of their wildebeest. Hyenas, much faster and stronger than the small primate, frequently die in just this way. On the other hand, bands of hyenas are known to be able to drive a lion from its meal. It seems safe to picture a gang of bipedal (perhaps rock-throwing) primates creating sufficient disturbance to distract a lion long enough for their purposes. Young Masai boys today, with a herd of cattle, are equipped to deter lions with only a spear and a long stick.

Eventually, our ancestors learned to kill their own game. No doubt, for ages they had been able to catch certain birds and reptiles, and the slowest of mammals, but this must have been on a very unpredictable basis. It was probably easier to catch middling to large animals than smaller ones. Alone in the woods or on the prairie, it would be virtually impossible to snare a rabbit, or even a mouse, unless one had a well-developed technique. It would be much easier to stalk a herd of antelopes and sneak up on a newly born calf. One may speculate that this was how the killing of larger animals began – by culling out the helpless young and aged. But as soon as australopithecines began to go hunting in parties, the situation changed drastically.

Much attention has been given recently to the social behavior – including the hunting techniques – of carnivores which hunt in parties, such as lions, wolves, hyenas, and Cape hunting dogs. In their provocative paper, "The Relevance of Carnivore Behavior to the Study of Early Hominids," George B. Schaller and Gordon R. Lowther, while emphasizing that they are not attempting to reconstruct the social life of early man, suggest that "the study of the social carnivores offers numerous possibilities towards the elucidation of the origins and form of social organization in man."

Early man, unlike all other primates, was an organized hunter. It behooves us, therefore, to look to those animals which are social hunters. Thus the *ecology* of early man becomes more important than his primate genealogy. His (genetic) primateness gave man his general physical attributes, but his way of life made him totally different from any of his primate relatives.

There are dangers in big-game hunting, even though the target animal may not seem particularly fearsome. Many large African antelopes have extremely efficient horns and hooves, and even the biggest and strongest carnivores sometimes have difficulties with them. Some of the smaller prey species, too, can be hazardous to tackle. The truculent wart-hog is one of these. The film "Born Free" portrays an inexperienced lioness chasing and bowling over a medium-sized wart-hog which, far from giving up the contest, runs straight at the big cat and strikes her in the ribs.

I once followed a pack of Cape hunting dogs on a late afternoon run over the Serengeti Plain. At considerable risk to Land Rover, photographic equipment, and personnel, we managed to keep up with the smoothly moving skein of dogs until they split up into two groups. I kept up with one bunch, who soon took off at a fast run after a large male wart-hog. After a considerable chase, the pig decided he had had enough running, and turned to face his four or five pursuers. The dogs, apparently perplexed, ringed the boar at a respectful distance and each gave the impression of waiting for another to be the first to risk the razor-honed tushes. A couple of the dogs sat down; one lay down. Finally, what I can only describe in an anthropomorphic way as a fatalistic "shrug" seemed to pass between the dogs, who turned and trotted away. These stand-offs may not be rare. The dogs usually kill by disemboweling their victim, and frequently most of the initial damage is done while the animal is still in full flight, with the dogs nipping and slicing at it from the side and rear. An animal which stands its ground may very often live to run another day. The same holds true with wolves; David Mech says that a bull moose can stand against a pack of fifteen wolves, but if he turns and runs, he will surely be attacked.

The cheetah, which usually hunts alone or in a very small group consisting of a female and her cubs, depends on sheer speed over a limited course. The blinding yellow blur of a cheetah in full run is one of the most stirring and heart-stopping sights a naturalist can experience. The animal begins by stalking a herd of gazelles, periodically "pointing" with a natural elegance no in-bred, pampered hunting dog can remotely approach, until the distance is a quarter mile or less. Then, selecting one individual animal as the target, the cheetah takes off – at first with long smooth bounds which make it very conspicuous (stealth is no longer necessary). Then, gradually quickening the pace, it appears to "shift gears." For the last hundred yards or so, the cheetah moves with amazing speed – at least sixty miles per hour. In the last bound of the chase the cheetah may use its foreleg to trip its quarry up or bowl it over. I have looked at film over and over again in an attempt to see the "moment of truth," but without success. After the cloud of dust settles, the cheetah will be lying beside its prey or over it, apparently giving the *coup de grace* by strangulation.

It is a killing pace, and if the cheetah cannot catch its target within a quarter mile or even substantially less, it will usually give up. Even when the chase has been successful, the cat is almost invariably winded. In the course of one kill, two small cubs which had been left far behind during the long sprint had time to walk up to their mother, open the belly of the dead gazelle and begin feeding before the old animal could get her wind back. (In this case, a hyena came along just a few moments later, drove the cheetahs away, and made off with their meal.)

Lions are even more short-winded than cheetahs, and must make their charges of thirty yards, more or less, from ambush. If they miss at that distance, they must hunt again. (It is surprising how relatively inefficient natural predators are. Misses are more common than successes, no matter how specialized the predator may be.) Except for loners, who must have a certain amount of difficulty sustaining themselves, lions hunt cooperatively. Most of the killing is done by the females, although males are perfectly capable of catching their own food if they are forced into it.

The lionesses often use a group technique which involves one animal frightening the selected antelope, zebra or other victim toward the others, which are fanned out in hiding.

Large cubs are frequently allowed (one might say "en-couraged") to participate, although they take a long time to develop their skills to the point of being able to take care of themselves. Sometimes a party of grown cubs will swarm over a large animal such as a buffalo, raking it badly but doing little real damage, until their mothers and aunts finish the process. Lions kill with their jaws, by grabbing an animal by nose or throat, and suffocating it. Indian tigresses have been known to repeatedly bowl over large animals such as water buffalo, apparently so their yearling cubs could get the feel of dealing with large prey; they use much the same killing technique as the lion.

At a lion kill, the adult male or males always eat first, regardless of the fact that they did not usually provide the meal themselves. They are followed – only at their lordships' pleasure – by the females. Finally the cubs are permitted to approach and find what scraps they can, but only after the adults have gorged to the point of exhaustion. Filled with wildebeest or zebra meat, the lions pant wheezily, lying wherever and in whatever position they happened to fall, bellies distended, paws waving limply in the air.

In times of food shortage, such as when the large herds of migrating hoofed animals are absent from their hunting area, the feeding hierarchy of the lions can be hard on small kittens and growing cubs. Then, the smallest members of a pride are so gaunt that their ribs are plainly visible. If the adults are not able to make a substantial kill very soon, these youngsters will fall behind and no doubt be picked up by hyenas or even leopards. This is a drastic but undeniably efficient method of adjusting lion populations to the available food supply.

Undernourishment is not as grave a problem for cheetah or leopard cubs because these animals do not consort together in social groups, and there are no additional mouths to feed after mother and young are satisfied. Also, both spe-cies tend to hunt in their immediate vicinity, and the cubs are rarely far from the scene of action. Cape hunting dogs, on the other hand, may run for several miles in the course of a chase.

This could make matters difficult for those members of the pack which are unable to keep up, such as small young and their nursing mothers. The dogs demonstrate attractive social grace by bringing food back to their dens. They do not carry pieces of meat in their mouths; they swallow them and regurgitate the meal when they arrive home. Wolves behave similarly. However, no cat is known to do this.

Despite stories to the contrary that one often hears from African farmers and hunters, hunting dogs are friendly, sociable, and gentle. White settlers have long told the most horrendous yarns about the animals (precisely the same sort of fiction one hears about wolves in other parts of the world), always with the emphasis on their ruthless viciousness. There appears to be no basis for these stories. In fact, the van Lawick-Goodalls call hunting dogs "Innocent Killers" in their recent book, which sheds an entirely different light on the animals. (The old-time Canadian newspaperman, Jim Curran, used to say, "Any man who says he's been et by a wolf is a liar.") The word "altruism" was coined by August Compte in 1853; if the capacity did not exist in man prior to that time it most certainly did in wolves and other dogs. If there can be kindness and amiability in the nonhuman world, then both are exemplified in these engaging creatures.

We usually think of killing in terms of aggression against one's own species. When we recognize the sort of killing implicit in predator/prey relationships, we view it in terms of lust, rage, and "animal" ferocity. However, at the moment of attack a carnivore is anything but angry, anything but "fierce." If there is a common aspect to these situations, it is the look of intense *alertness* about the animals. Ears are pricked forward, not laid back as in anger; eyes are wide as in anticipation, not narrowed; tails are raised as though in pleasure, not dropped as in stress; jaws are fairly "smiling." You may see a snarl if you approach a cat's meal or its young, but not during the hunt. One need only watch a house cat hunting butterflies or grasshoppers in the garden, or chasing a ball. He is obviously having fun, and his expression is the same as that of his large relatives at the moment of the kill. At that instant the animals are doing what they do best.

Proto-men and early men were not as swift as the carnivores who shared their African hunting grounds, nor were they as strong or well armed. However, they were organized. Social or group techniques of hunting, such as those of lions, make up for the animals' limited attacking range. Sheer numbers in the case of the relatively small hunting dogs, or wolves, compensate for their comparatively minor armament. Similarly, man-apes and early men could offset their physical limitations with both numbers and organization. The "drive" could be undertaken in a dog-like way, and the ambush could be carried out with the techniques of the lion. Thus it is that lions and dogs – social carnivores – may tell us more about early man than our nearest living primate relatives are able to.

In their earliest hunting encounters with big game, our ancestors would have been mainly on the watch for animals which were at a disadvantage – an unprotected fawn, or an aged, wounded or otherwise crippled adult. They were probably quick to take advantage of animals mired in mud holes. There comes to mind one aged female Cape buffalo, blind in one eye, which had become stuck up to the withers in a mud wallow. There was no possibility of escape for it. By the next morning, hyenas had consumed the defenseless creature.

George Adamson tells of the periodic droughts which occur in the Lorian Swamp of eastern Kenya. One year, at the center of the swamp, "there was a pool of semi-liquid mud containing a score or more of rotting carcasses and in the midst a seething mass of dead and dying cat-fish. Further on we came on the bodies of two cow elephants and a little calf embedded in the drying mud. Next to them was a camel cemented into dry mud." His description is reminiscent of the tar pits of Rancho la Brea in California, which have provided a bonanza for the vertebrate paleontologists. There, thousands of creatures of many species became mired in the tar. One would imagine that predators and scavengers could themselves be trapped in such situations, but the elderly buffalo in a small mud wallow presented a perfect opportunity without risk.

For a social animal in the process of developing a remarkable brain, it was probably not a giant leap of the intellect to think of driving prey animals into natural traps such as bogs and soft muddy water holes, or into *cul-de-sacs* where they could be dispatched with stones, large rocks, and eventually "spears" which may have been sturdy green sticks. The simplest method was to drive herds of animals over cliffs – until recently a standard technique for killing bison in North America. The waste must have been colossal, for the amount of meat which even a large party of hunters could consume or carry away was limited.

Social hunting and scavenging involve sharing of the spoils. The behavior of lions in their rigidly hierarchical table manners – adult males, adult females, young – certainly cannot be considered social. Their hunting technique is cooperative, but when dealing with the booty, the lions exhibit "selfishness." Schaller and Lowther suggest that the evolution of the social system of lions is "incomplete in that it includes cooperation in hunting but not in the sharing of the prey."

By our standards, a lion kill is mild when compared with the almost orgiastic mayhem of wild dogs and hyenas. When the hyenas make a kill – often on a moonlit night in a group which may number two dozen or more – they tear and bolt their food with an urgency that must be seen (and heard) to be appreciated. Yet, despite the astonishing speed and greediness with which they devour their spoils, the hyenas and dogs do not really squabble over it. Every member of the pack seems to get its share, and they transport food back to those unable to take part in the hunt.

Kills of relatively large game inevitably attract other meat-eaters. Lion kills attract vultures, jackals and hyenas; and hyena kills attract lions. Australopithecine kills must have called attention to themselves in the same way, and no doubt the larger carnivores would frequently drive the small primate hunters from their prey. It follows that the latter would make off with whatever they could well before darkness. Like the hunting dogs, they probably carried food back to their females and young.

It is difficult to determine where *Australopithecus* had his home base. We customarily think in terms of caves. Caves were used by men of the genus *Homo* much later on, but it was probably not the best place for a small primate until he had developed exceptionally good weapons. Perhaps fire and associated amenities preceded cave-dwelling. One might guess that ancient hunting bands congregated at night under an overhang on a rock pile, or in a log-jam of treetrunks overturned by elephants, or perhaps in the trees themselves.

Food-sharing must have eventually contributed to the evolution of work-sharing. The mature males were the active hunters; and the females and young – or at least those females presently with small young – did not participate in the hunt. One can assume that the centrally based females gathered vegetative material and perhaps minor animal food, but the care of their infants prevented them from taking part in big-game hunting.

As the australopithecines grew larger, their young became increasingly helpless; and with the process of evolution, they became more infantile at birth, more "premature." As the size of the skull increased, it became more difficult for it to pass through the female pelvis. The solution was to bear the young at an earlier stage of development, which would then allow the head to move through the pelvis. Prematurity at birth meant longer infancy, and a longer period of infancy resulted in more time in which to learn. In a feedback process, the brain grew faster, and young were increasingly helpless at birth. More energy and time had to be expended in their care, which led to new patterns of social behavior.

For modern man, growing up is a slow process. We are in the uterus for nine months, we have our first permanent teeth at about 6 years, female reproductive period begins at about

15, and we have our last permanent teeth at about 20. A chimpanzee, by contrast, matures at a dazzling speed: gestation is about the same, but first permanent teeth appear at 3, female reproductivity at 8, and last permanent teeth arrive at about 10. No doubt early man fell somewhere in between. Young proto-men were almost as vulnerable as human infants, which means that the social conditions of their lives must have been elaborately developed for the early communities to survive.

The sharing of the work-load, with the adult males hunting and the adult females guarding their young, suggests that there were un-primatelike sexual conventions by this time. Since these were cooperative hunters, it is unlikely that one dominant male copulated with all females. There may have been proper "pair-bonds" such as we see among birds. These bonds would be able to tolerate the temporary absence of the male hunters. The moment a dominant or "Alpha" male baboon's back is turned, one of his females in oestrus (heat) will be approached by one of his adjutants. For a hunting society to hold together, the baboon way of life would not be appropriate.

The emergence of year-round female sexual receptivity enabled evolving man to deal with the problem of periodic male absences. The female would thus be continually attractive to her mate. In all female primates save man, sexual receptivity occurs only in periods of oestrus, the time of which varies from species to species. When she is not in oestrus, the female does not entertain male overtures, nor is she attractive to the males. The moment she does come on heat, she may be the only individual in that condition at the time, and in many primate species she may mate indiscriminately with as many males as there are in the band, or as many as can get away with it under the watchful eye of "Alpha." (Schaller tells us that gorillas show no sexual jealousy whatever.) In early man, with all the females receptive the year round, it was no longer necessary for the males to watch for the currently receptive female. Lasting sexual bonds could be forged between individual pairs, bonds which were strongly reinforced by the constant presence of infantile young. The monogamous pair with young was now the family. (Carl O. Sauer has suggested that the weakening and loss of the oestrus cycle "is probably a feature of domestication, and it may have occurred early in the history of man, eldest of the domesticated creatures.")

No doubt several such families preyed – and stayed – together. The helplessness of newborn young and the relative helplessness of growing young for periods which now extended into several years meant that defensive mechanisms were well developed by this time. Certainly there was communication of some kind, perhaps "primitive" by our standards, but still the most sophisticated that had yet evolved. Communication does not need to be very elaborate for warning and alarm; in those circumstances the simplest symbol can convey the meaning. But since these were highly social, cooperative hunters, the defense of their home sites must have been organized and effective.

By the time these initial communities had developed, there must have begun to emerge a subtle or even a mildly conspicuous change in the attitudes of proto-men toward their surroundings. Confidence must have been burgeoning. The effective hunter and the effective defender of his family had probably begun to feel a change between himself and the rest of nature. Proto-man could probably not articulate this thought, nor perhaps even entertain it consciously, but the *feeling* was there. The primate place in nature was changing, to a degree never before experienced in the history of any form of life. A creature that could do all the things the australopithecines were doing was also a creature which had sufficient intelligence to sense what was happening – that "nature" was a lessening threat and an increasingly dependable source of the family welfare.

We shall never know when *consciousness* of this change arose. We can only speculate. It is unavoidable, however, to conclude that the social hunter felt increasingly confident about his prowess and about the security of his home base. It is too early to call home base a campsite, probably, because it was informal and very likely shifted position more or less regularly, but it *was* a base. At some stage, the hunters' systems of defense became so efficient that newer and even more un-primatelike modes of behavior emerged.

One of the most notable of these new forms of behavior was face-to-face (or belly-to-belly) copulation, in the style sometimes known today as the "missionary position." All primates save man habitually copulate in the usual manner of quadrupeds, with the male approaching from the rear. (The female, if she is not receptive, can simply walk out from under the male.) Some fascinating theories have been advanced regarding the contribution the new sex position may have made to social behavior in general. Certainly it was convenient for talking – or at least for some kind of oral or facial communication – and presumably it made sex a more ritualized and thus perhaps a more important element of social life than it had been in the past. It may have represented the dawn of affection. Some authors suggest that the female, receiving her mate face to face, might have responded emotionally as she does to the infant at her breast. This may have given rise, according to Hockett and Ascher, to "such further consequences as the Oedipus complex."

Ventral copulation was made possible (but not absolutely necessary) by bipedalism. The buttocks having developed to such a great extent as part of the equipment for an erect stance, and the vagina having shifted from a horizontal to a vertical position as the immediate result of standing upright, in addition to any emotional benefits that may have accrued to it, the new position was simpler mechanically. It is also suggested that the euphemism "sleeping with" rather than "copulating with" arose not in prudery but in the fact that the supine female, with her vagina restored to the ancestral horizontal position, has a better chance of retaining semen than it would if she were standing up or running around. Also, the female orgasm and post-orgasmic relaxation and lethargy could induce her to remain supine for an additional length of time.

More important, belly-to-belly, horizontal sex may well reflect evolving attitudes not only in the human social context but also with surrounding nature. It is difficult to imagine a female primate allowing herself to be pinned on her back by a heavy partner when every surrounding tree or rock might conceal a hungry leopard. Animals simply do not like to be tipped on their backs, and they especially resent being held in that position. It makes them far too vulnerable. Before the new position could become a firm tradition (after all, dorsal copulation is still entirely feasible for man, and widely practiced) there had to be a profound change in attitude. There had to be security from potential predators. The female had to feel especially safe, not only for herself but for her young, before she would have countenanced anything as "unnatural" as this.

It meant a fundamental change in awareness; it might even have been a dawning of the "self" for the individual and, more important, for the species. It shows an indifference to the surroundings which is one of kind rather than of degree. The change was basic and radical. From that point onward, man was different from anything else in nature. Perhaps, in more ways than one, sex made us what we are today. When, as proto-men, we began to enjoy sex in the new position, the leopards did not attack us. They had already learned the meaning of fear.

Exogamy – mating outside the family group or the immediate "tribe" – must have arisen relatively early, if not as a conscious social institution at least as a behavioral pattern. We can be certain that the hunting man-apes knew nothing of incest nor of taboos, but as in other social carnivorean societies such as those of lions, the young animals after adolescence tend to scatter away and to establish prides, or families, of their own. The effect of this is to ensure a proper mix of lion genes, and no group becomes overly inbred. Man, the carnivorous primate, went in the lions' direction rather than remaining in the dominance hierarchy tradition of baboon societies. Washburn, the baboon authority, emphasizes that baboons have the opposite tendency. They breed constantly within their original group, with very little outside mixing. Baboons, of course, are not hunters. The way of life of a species is much more important to its behavior than its phylogenetic relationships.

Some form of "tacit agreement" seems to underlie exogamy. The mature adults of many species apparently do not want young adults nearby any more than the young adults themselves want to remain close. It appears to suit everybody. It might be hard on the pair-bond if there were many distractions too close to home, and so it would appear that unlike that of nonhuman primate societies, the tendency to splinter off and form new groups was a social factor in earliest times. Initially, the reasons were biological, not cultural. Exogamy arose in a new species of primate living a new kind of life in order to maintain cohesiveness in the structure of a carnivorean family.

Chester S. Chard calls our attention to an additional very interesting point made by Bernard Campbell, who suggests that the very rapid evolution of our forefathers "may prove to be correlated with the appearance of exogamy." Campbell's point is that exogamy would result in a substantial increase in genetic variability "in contrast to inbreeding primates with their slow evolutionary change over millions of years." The wider the mix, the greater the chances for evolutionary change, and the faster the process accelerates.

Assuming that young australopithecines of both sexes drifted away from their families to mate with others of their age group, we must now face the question of relationships between these families and other groups. Did they fight among themselves, or did they share their hunting grounds? Was the pre-human hunter territorial?

The territories which animals defend against others of their species tend to space out populations and to parcel out available food resources in an equitable way, so that there is not too great a strain on those resources in any particular place. Animals vary widely in the extent and intensity of their territoriality, from fiercely competitive little fishes and backyard birds to the more *laissez faire* attitude of the Cape hunting dogs. There are few general rules about territory in the mammals. Some mammals are vigorously territorial, some are not. As with other social mores, it is more a matter of environmental context and feeding habits than of phylogenetic relationships.

Thus, in seeking evidence of territoriality in the hunting man-apes, we are advised to look not so much at the behavior of present-day primates (which exhibit varying degrees of the custom but among which our closest relatives, the great apes, show little) as to consider the territorial habits of social carnivores. Here, there is equal confusion. The Cape hunting dogs, exemplary practitioners of food-sharing, do not appear to be strongly territorial. They range wherever the hunt takes them, except at the pupping season, when they tend to stay reasonably close to the lactating females with small young. The puppies do not take long to grow sufficiently large to run with the pack, and there would seem to be little need for the dogs to maintain or defend hunting grounds.

Hyenas, on the other hand, *are* territorial. Their hunting methods are much like those of the dogs: a pack of animals chases after a victim until it tires and is torn to pieces. Hyenas do not usually run as far as the dogs do, however, and they invariably return to their dens, which are located in traditional territories. Hans Kruuk, who pioneered studies of spotted hyenas and their behavior, discovered that in Tanzania's 102-square-mile Ngorongoro Crater the available land was divided up into eight (matriarchical) hyena "clans," and that a clan might contain as many as 100 individuals. The territorial boundaries change from time to time, but at any given time the boundaries of the moment are fiercely defended by the resident packs.

Lions fall somewhere in between the hunting dogs and the hyenas. They do identify territories, but on the other hand they are often very lenient about passers-through, and indeed some individuals and small groups seem to be confirmed

wanderers with no home base at all. Wolves, according to
David Mech, are probably territorial (they have been observed to attack and drive off lone interlopers), but if they
exist as such, the territories are so large that there is little
chance of two packs running into each other.

Into which category did the ancient hunter fall? Did he
stay put? Probably not, at least until his bands had grown to
substantial size. He must have moved around a good deal
more than the average lion does, but somewhat less than a
hunting dog, having in mind his physical limitations. There is
presently no evidence of territoriality in early man, but
neither is there anything to refute it.

It would seem reasonable to expect that there is the germ
of territoriality in all land mammals. Territoriality in man is
probably a "pre-adaptation" – a physical or behavioral characteristic which does not reveal itself until the need for it
arises. In man, territoriality would only be manifest in situations such as overpopulation in a given area. Perhaps as numbers of the early hunters grew, there developed in some
places traditionally defended core areas and hunting areas.

Territoriality is a phenomenon which exists for the purpose of spacing out animals of the same species, and has no
relation to other species which may inhabit the same stretch
of landscape. Territoriality exists between man and man,
hyena and hyena – not between man and hyena.

Man has been repeatedly charged with "innate aggressiveness." This aggressiveness is conspicuous both in our relations with other people and with other species. It raises the
question of dominance, whether it be over one's brother or
one's nonhuman cousin. This is the question of the "killer
ape."

In the minds of most people the word "killer" conjures up
a vision of diabolical fierceness and cruelty. Our ancestor was
not an ape that was a wanton killer, but an ape that was a
predator – a hunter – no less amiable than any other ape. Animals which are strongly territorial react strongly to others of
their species who invade their personal plots. A male bluebird
will vigorously chase away another male bluebird from its

wooden fence-post. That does not make the bluebird a fiercely aggressive creature.

We confuse aggression with predation. Predation is a method of food gathering. Is there much difference between picking up a bird's egg or a nestling? Or in taking an adult bird? Or a hare? Or a gazelle? These are only matters of degree. Those who call the pre-human hunter the "killer ape" are not only being misleading, they are offering a dangerous simplification.

We must think of the pre-human hunter as a social carnivore. The fact that he was a primate is incidental. The great apes are remarkable for their gentle dispositions. We can quite properly assume that the last of our forebears who left the trees were not much different in the social sense than the apes we know today.

Once they became social carnivores, our ancestors were like no primates that ever lived before. Their way of life was completely new, for primates. Were the new hunting-and-gathering ape-men dangerous to other animals around them? To their food victims, they certainly were. However, when they were not hunting, the australopithecines were probably just as indifferent to their surroundings as a gorged leopard is today. It is difficult to conceive of a horde of "killer apes" launching ferocious attacks on anything that moved. Predators do not behave that way. Mayhem is a waste of energy, and energy is what the sleeping cat and dog are conserving.

There was no "killer ape," but there is a "killer man." Human aggressiveness has been *learned*. It has been transmitted through uncountable human generations. But it did not originate in the trees. It is not "instinct." It arose in the social and cultural context some time between the day when the first ground-dwelling primate picked up a weapon and the day when man became human. Since aggression is not instinctive, but is learned, we may have hope. Cultures and traditions can change, and they can change rapidly. Perhaps a new culture will abandon our destructive traditions, realizing that these traditions are not "animal." Perhaps we will recognize that we are one with the animals upon which we prey.

CHAPTER EIGHT

Fire, Ice and Megadeath

The Pleistocene is generally associated with glaciation. But there is another reason for intense scientific interest in this strange period. The latter part of the Pleistocene (up to about 12,000 or so years ago) saw the extinction, *without replacement,* of large numbers of species of large mammals – all over the world. This was especially conspicuous in North America, where a long list of the very largest mammals became extinct, according to the evidence, at about the same time as man was establishing himself in the "New World."

This phenomenon has been described as "Pleistocene overkill" by Paul S. Martin. The extinctions occurred on a world-wide basis, but the fact that they were so pronounced in North America has led to much questioning and controversy. Extinction has always been the natural way. Extinction is the inevitable future of all species, genera, and even faunas. But extinction has always been gradual, involving the replacement of the extinct forms with new ones. However, the large mammals of the Pleistocene that disappeared did so without descendants and without the emergence of new faunas to succeed them.

Immanuel Velikovsky, the controversial author of such books as *Worlds in Collision* and *Earth in Upheaval,* maintains that there have been cataclysmic upheavals going on in the world all along. These persisted even into historic times, long after the departure of the Pleistocene mammals. Sudden, without warning, and gigantic in scope, these changes engulfed whole populations of animals and brought about the extinction of such creatures as woolly mammoths.

It is true that some of these animals must have met with natural misadventure. Their bodies have been found frozen in the ice, complete with greenstuff in their stomachs. When a bank of the Beresovka River in eastern Siberia collapsed and fell away some years ago, it exposed the fresh-frozen remains of a mastodon. Thawed, it provided "fresh" meat for hungry sledge dogs, which seemed none the worse for downing a meal which had been waiting for them for 20,000 years. Mastodon flesh has even been served at scientific banquets (in very small quantities). The consensus is "only fair." However, one does not need to subscribe to Velikovsky's cataclysmic theories to acknowledge that animals are frequently overtaken by such misadventures as blizzards of unusual intensity. This happens from time to time to bison and domestic livestock on the northern Great Plains during especially severe winters. There is no reason why it should not have happened from time to time to the northern elephants, but it is difficult to accept this as the cause of their total and complete extinction, especially over such a short period of time.

Climate must have played a part in the extinctions of the Pleistocene, as it has through all the ages. But it could not have been climate alone. There was an additional factor, and that was man. It is known that man hunted and killed mastodons and mammoths. For years, naturalists have entertained the notion that some of these great elephants of the Pleistocene may still be wandering the more remote Siberian forests. A more credible thought is raised by one scientist who believes that, in view of the fate of African elephants at the hands of ivory poachers, the last Siberian mammoth may have met its end for the needs of the Chinese ivory market.

The magnitude of Pleistocene overkill was appalling. The record shows an extraordinary acceleration in the rate of extinctions without replacement. Accelerations in the rate of extinctions usually involve some kind of *Homo* in the woodpile. It seems to be a sad fact that when animals disappear on the grand scale (as opposed to the gradual changes of the natural evolutionary scale) man is almost always implicated. Now, human impact on other species moves back into prehistory.

In *Wildlife Crisis,* James Fisher summarized some of the recent extinctions on various continents at about the time of the development of more sophisticated human hunting techniques. In Britain, "man has been a major influence on ecological, faunal and floral changes . . . for 30,000 years at least." In the period between 50,000 and 10,000 years ago, Fisher notes the last British representatives of the cave bear, leopard, forest elephant, sabre-tooth, woolly rhino, lion, and woolly elephant. He assigns culpability for "the demise of some, perhaps most, of the larger mammals" to human hunters.

In the Middle East, a famous deposit of human bone fragments and artifacts, and animal remains, has been uncovered in the Central Jordan Valley near the Sea of Galilee. There are the bones of "some thirty-seven species of mammals, at least thirty of which are extinct." With regard to Africa, Fisher quotes C. D. Darlington: "In Africa 100,000 years ago men . . . exterminated many genera of great mammals which they killed for food." Martin, proponent of the overkill theory, remarks that despite the wide array of large animals in Africa today, they represent only about 70 per cent of the genera which were there in the middle of the Pleistocene. Stone Age man appears to be inextricably implicated. Simplification of natural systems, including simplification in terms of numbers of kinds of animals, does not happen naturally.

Man is thought to have been in Australia for about 30,000 years. Although Australia's indigenous mammals are all marsupials, there were some very large forms extant at the time of the human invasion of the continent. All of the largest of these are now extinct. There are thought to have been 27 different species of large, flightless moas in New Zealand when the first Polynesians arrived there; none remains today. There are similar accounts from other oceanic islands of the world, and the evidence linking human occupancy to the extinctions of so many of the largest creatures appears to be overwhelming.

The most obvious overkill was that which occurred in North America. Man is a rather recent arrival, in terms of natural chronology. One might think that since man originally evolved on the African plains together with the large animals which still live there, the latter may have been in some kind of equilibrium in a natural way. But the large animals of North America had been totally ignorant of man from their very beginnings, and had had no "conditioning" prior to the invasion from Siberia. It is possible that the arrival of a new kind of predator – especially a technological one – was too much for them to handle. Man critically altered the natural equilibrium. In North America close to one hundred mammals and birds have become extinct – without replacement – in the last 32,000 years or so.

At various times during the Pleistocene there were periods when, because of glaciation, much of Earth's waters were taken out of circulation with the result that sea levels fell substantially. During these periods, the Bering land-bridge was exposed, and there was free movement of animals back and forth, in both directions. At various times horses and camels, which evolved in the Americas, moved westward into Eurasia, and the ancestors of moose, caribou, bison and others moved eastward. With the latter came man.

It has been extremely difficult to determine the timing and the progress of the human invasion of North America. Fossil and archaeological discoveries have been rare, or absent altogether. However, on the gradually accumulating evidence of stone tools found in the Yukon and in Alaska, and across the Bering Strait in Siberia, there has been for some time a credible inference that man came across the land-bridge more than 20,000 years ago. It would now appear that it was much earlier than that.

In 1961, A. Mac S. Stalker, of the Geological Survey of Canada, found human bones on the east bank of the Oldman River near Taber, in southern Alberta. They were the badly broken-up skeletal remains of a child less than two years old. It has been impossible to perform radio-carbon dating on these fragments, because they are too small, but Stalker's 1969 estimate of their age, based on the geological evidence at the site where they were found, is upwards of 37,000 years. In his opinion, "they are considerably older than that, and perhaps as old as 60,000 years."

The investigations of most archaeologists and palaeon-tologists has been drawn considerably farther northward. At various intervals in the glaciation, parts of Alaska, the Yukon, the Mackenzie delta and valley, northeastern British Columbia and northern Alberta were free of ice, and a "corridor" existed which allowed penetration to the interior of the continent. One would assume that these ice-free areas (*refugia*) contained plenty of big-game animals for the late Stone Age hunters. The assumption is that the nomads from the west moved across the top of the mainland to the region of the Mackenzie delta, then southward down the ice-free valley toward the interior. Thus it is possible that, far from being deterred by the last maximum glaciation, men moved south to the Great Plains in the very middle of it, due to the physiography of the time. Indeed it is probable that there were repeated invasions from Siberia over many thousands of years.

We must not think of the human invasion of North America as consisting of a host of marching men, sweeping southward like so many Assyrian cohorts. More likely it was gradual, low in volume, but a trickle which never really stopped. No doubt the hunters followed the movements of the large mammals they depended upon, such as caribou, bison, moose and others. They probably kept up with their seasonal supply of food in much the same way as wolves follow caribou today, or hyenas follow wildebeest.

Some of them, however, give the appearance of having moved rather quickly. Human sites at the extreme southern tip of South America (Tierra del Fuego) have been dated at more than 10,000 years ago. That was a trip of some 11,000 miles from the Alaskan coast, if indeed that is how they got there. There are human dates of 14,000 years more or less from Venezuela, and slightly under 10,000 from Brazil. Fisher remarks, "In the cave faunas of Lagoa Santa in Brazil and in several caves in southern South America is evidence of a vast pampas Pleistocene-type assemblage of big game that could have competed as a show of specialized monsters on even terms with any other fauna of the time, including that of

Africa." Certainly this assemblage is not there today, and has not been there in historic times.

A great deal of the evidence in the question of Pleistocene overkill has been circumstantial because of the difficulty of finding human weapons in incontrovertible relationship with the remains of now-extinct large forms of wildlife. The evidence is still being accumulated, but some of it is no longer circumstantial. In the Old Crow area of the Yukon Territory, for example, scientists have discovered bones of the extinct mammoth, broken while fresh, together with human tools, in this case a kind of flesh-scraper fashioned from the leg bone of a caribou or reindeer. Other extinct species found in the same area, together with evidence of human occupancy, were horse, camel, mastodon, ancient bison and a kind of lion, among many others. Radio-carbon dates for the mammoth bones and the fleshing tool coincide at between (roughly) 25,000 and 30,000 years ago.

The Yukon discovery fits neatly with the geological estimate of an ice-free corridor in that part of the north between 35,000 and 25,000 years ago. (Recent evidence of human presence in Mexico some 24,000 years ago, however, together with the southern Alberta material, may push the chronology back considerably farther in the near future.) No one suggests that human invasions may not have taken place in an even earlier ice-free period when the Bering bridge was also present, but the evidence for that has not yet been forthcoming.

Proponents of the Pleistocene overkill theory of large mammal extinction show that the zenith of large game exterminations seems to have occurred between 10,000 and 15,000 years ago, with 11,000 years being the critical period (which, coincidentally, was the date of the beginning of the last retreat of the ice). This is also the rough date of the disappearance of the mammoth, although some did persist somewhat longer than that. The last mastodon may have perished two to three thousand years later; Paul Martin is skeptical of dates more recent than that.

The issue of man's role in the extinction of the great Ice Age mammals is still being vigorously debated. However, it would seem that the disappearance of so many of the largest and most specialized animals concurrently with the period of human penetration into almost all parts of the world is significant. The chain of events probably involved massive kills of the more slow-moving and easily trapped herbivorous animals. Their disappearance made it possible for lesser forms to increase in the vacated habitats, but it also drastically cut down on the numbers of carnivores such as sabre-tooths and giant jaguars which depended on the larger species for food.

A crucial consideration in this investigation involves the time-scales concerned. Because we, in our present form (the Stone Age hunters were men in our present form) are such recent arrivals, we have a tendency to ignore the slowness of biological evolution, as compared to cultural evolution. In *Pleistocene Extinctions: The Search for a Cause,* W. E. Edwards makes this point most persuasively. He emphasizes that for hundreds of thousands of years, the evolution of human culture did not have any appreciable effect on the numbers of animals that men hunted. The australopithecines, and later the "dawn men," were gradually becoming more efficient at hunting, and slowly, almost imperceptibly, they took over from the more traditional scavengers and predators such as the big cats, dogs, hyenas. But – and here is the critical point – "like a population growth curve, the curve of cultural evolution rises exponentially." Once it got under way, it is probable that no adaptation in nature ever evolved so quickly as human culture and hunting efficiency. Biological adaptations were unable to occur and become established fast enough to offset or compensate for the cultural trend. In ancient times, things had progressed in a more or less parallel fashion: as the hunter became swifter, so did the hunted, and as the hunter became more heavily armed, so the hunted became stronger. Not so, however, with evolving culture. The prey species could not keep up with it; genetic mutation is too slow.

Edwards observes, "In the New World the effect was virtually instantaneous." The large faunas of the Americas could not become adapted to cope with a predator that arrived with an already well-developed method of hunting. There was no way for biology to keep pace. There the evidence rests. Those large animals that did survive the Pleistocene overkill were those that then (and now) lived on inaccessible mountains, remote arctic islands, and so forth. Some of them, such as caribou, were those who knew man from time immemorial, and invaded North America with him.

There has been much speculation whether early man in North America (and elsewhere during the same period) made use of fire in driving prey animals to bay. There is little evidence of this prior to the time of the Indians. Although men had had fire, and used it, for hundreds of thousands of years before that time, it is not known to what *extent* they used it. Fire was being used by men long before the appearance of *Homo sapiens.* It has long been known that *Homo erectus* had fire, probably for many, many millennia. In fact, it has been seriously proposed that *Australopithecus* knew and made use of fire, but this evidence is under incendiary debate. We may never know when the use of fire became widespread, much less when the transition was made from the *use* of fire to the *making* of fire, which is quite a different thing.

Many animals are clearly fascinated by fire, and by smoke. And – strangely – by ants. The connection between fire and ants is not as tenuous as it may seem. Anyone who has been bitten by fire ants will know that each nip is as though a tiny burning coal had been pressed against one's skin.

Some animals seem to deliberately invite the attentions of fiery ants. A naturalist knows the peculiar habit of birds which is called "anting." A bird – robin, jay, thrasher, blackbird, oriole, among others – will, on encountering an ant hill, seize individual insects in its bill and insert them at various places in its plumage. It rubs the ants under the wings, over the back and tail, and in all the places one would scrub and sponge while taking a shower. Various explanations for the phenomenon have been offered, such as the possibility that the formic acid of ants may have insecticidal properties and thus help to keep body parasites under control. An alternative suggestion might be that all animals like to scratch and be scratched, and that ants may well produce that satisfaction for birds.

"Anting" sometimes becomes "firing." When birds cannot find ants, they will use such things as cigarette butts, burning matches, hot coals, and smoke. It would seem to be risky, but apparently the attraction outweighs any fear the bird may have of smoldering objects. The fascination with smoke and flame does not stop with birds. Much closer to home, there is in our own primate order the small tarsier of the Philippines which has been named *Tarsier carbonarius* for its habit of "picking up hot embers from campfire sites." Kenneth P. Oakley has observed that "this suggests that man's ancestors far below hominid level of evolution may have been attracted to natural fires and toyed with burning matter." He adds, "After all, is not tobacco-smoking – addiction to which apes and men are equally liable – a form of 'anting' behavior?"

Our ancestors probably had their initial acquaintance with coals, smoke and flames from natural fires. The Rift Valley of East Africa, where so many early proto-human remains have been found, is still a "hot" area in a volcanic sense, and may well have been more active several millions of years ago. Undoubtedly, volcanic eruptions were observed from time to time. Presumably also, once their initial alarm had abated, the observers would have been curiously attracted to small, smoldering patches of grass and dry shrubs around the edges of lava flows. And like the "anting" bird or the ember-oriented tarsier, the man-apes would not have been able to resist the new curiosity. Volcanic eruptions, however, are not all that frequent, and one finds it difficult to imagine that this could have been the sole origin of the tradition which, above all others, was to help man become the world's first technical animal.

On a world basis, lightning is much more common than volcanic eruption. About ten thousand thunderstorms take place every day, all over the globe, and each of these generates an average of ten lightning strikes. A total of one hundred thousand lightning bolts per day makes the chance of natural fire from this source a great deal more likely than from volcanoes. Of course much depends on where it strikes. A bolt in the drier parts of the boreal and mixed forest in August is much more likely to start a fire than one which strikes the grassland. Grass has always been thought to be extremely difficult to ignite in this way. However, since our antecedents evolved on the savannahs of Africa, it is there that we should look for lightning fires.

Open grassland was the habitat of so many different species of large grazing animals in Africa which evolved under those conditions that the grass must have been there for millions of years. African people know the values of firing the grass, because it produces fresh green shoots for their domestic animals. Before there were African people to ignite the grass, however, there must have been a sufficient number of natural fires to keep the prairie open. Otherwise, by this time they would have been covered in brush and forest and the grazing animals would long since have vanished.

Grass is extraordinarily resistant to fire; shrubs and trees are not. Something kept those savannahs open for millions of years, in order for the grassland-adapted animals to evolve. No doubt in the earliest times it was due to natural nonhuman agency. Somewhat later, one might guess that early man or even proto-man might have done his share.

Proof of the use of fire prior to the time of *Homo erectus* may never be forthcoming, but it is believed to have been in widespread use (whether purposefully or not) long before that. Cooking may have arisen quite early, as a form of partial pre-digestion of raw meat. Though our primate innards are quite generalized, they lean more toward the digestion of fruit and other vegetable matter, and easily absorbed material such as eggs, than they do to the long tough fibres of meat. One can speculate that before the "invention" or discovery of cooking, much of the meat consumed was probably a good deal "higher" than even the Englishman's pheasant decomposing on its string.

Fire must have contributed immeasurably to the burgeoning self-confidence of our man-ape ancestors and to the evolving sense of separateness from nature. Many animals use fire in one way or another, and many are not especially afraid of it, but man is the only animal who makes fire and controls it. Fire must have been an important morale-builder, although at no time since he became human did man's psyche need much reinforcement with regard to his relationship with the rest of nature.

There appears to be nothing genetic in the tradition of self-confidence which so quickly evolved into arrogance. Like other primates, man is very low on the "instinctual" scale, and has to learn almost everything he does. The cultural transmission from generation to generation of refinements in toolmaking techniques, of improved cooperative hunting methods, and of evolving social conventions of all kinds, was the essential underpinning of the conceptual man/nature dichotomy long before there was *Homo sapiens,* and quite probably before there was *Homo.*

By the time modern man penetrated all the continents during the Pleistocene only 20,000 years ago, his "dominion" over nature was well established.

CHAPTER NINE

The Power and the Glory

In the early days of modern man, before the carnage of the Pleistocene, weapons were simple and awkward to handle. Yet these crude weapons were capable of inflicting severe damage on even the largest game. One can visualize an exhausted, mired or injured horse, camel, or elephant being slowly crushed by rocks; or bludgeoned, often for a period of hours. One can also imagine sharp sticks being jabbed into a helpless beast, until it died from loss of blood. These proto-spears probably worked in much the same way as the giant fangs of the sabre-toothed cat, which could stab and slash through even the thickest skin of the large animals. No doubt the predator would often begin to eat its prey while it was still alive, as contemporary carnivores frequently do.

Even in those early times, as more sophisticated hunting methods evolved, it must have begun to dawn on early man that he, alone among the animals, was the weapon-wielder, the fire-user, the "dominant."

Although we shall never know when the sense of absolute power became an integral part of human assumption, we do have clues of when the need for power was *acted upon*. The important development was the change from sheer physical striving to *conscious* striving for power over animals. This involved insight, and it is difficult to imagine that men who had insight did not also have a concomitant sense of their status in their immediate habitat. They knew they could perform acts which would have predictable results, and they also knew that the prey animals were not privy to that knowledge. They could plan and execute a hunt much more efficiently than any pack of wolves or pride of lions could do, and they knew it.

Though one could not suggest that in proto-human times there was any human conception of "nature" as apart from man, the earliest *Homo sapiens* undoubtedly recognized his burgeoning power over other creatures. Every success contributed to his swelling confidence, and this in turn "fed back" into increasing efficiency.

Neanderthal practiced cannibalism, and occasionally buried his own dead. He even strewed fresh flowers about the deceased. Elephants and chimpanzees have also been known to place branches over bodies of their kind. This would seem to be a demonstration of affection which is not manifest in the lives of most animals. Neanderthal affection for the departed may be reflection on pleasant things past, or it may involve a form of conceptualizing about the future. One hesitates to speculate that this might include an intellectualization of "life after death," but it does indicate that care for other members of the Neanderthal group extended to the time of death, and – even briefly – beyond.

It is easy for us to say that Neanderthal burials were the first manifestations of consciousness of the finite time-span represented by an individual life. One would expect that this knowledge had long been a part of human culture. Men had seen countless lives – both their own and those of prey animals. The transitory nature of one individual career – whether bear, caribou, or man – must have been well known long before that knowledge was translated into ritualistic behavior. Other animals appreciate death, and other animals mourn. Neanderthal had been mourning for a long time before he performed the ritual burial.

Neanderthal must have "known" that he was not the same as the animals he preyed upon. Otherwise he would not have buried his dead. No other animal does that. Self-consciousness had begun to show. The difference between one's self and one's prey was so developed by this time that Neanderthal was able to conceptually transpose living animals to animal symbols, thus bringing the animals more and more under his control.

The symbolization of animals was the dawn of magic. It began when the Neanderthal hunter noticed a twisted stump or a peculiar rock formation which quite fortuitously resembled one of the animals he hunted. In conditions of bad visibility, he might even have imagined it *was* the living animal, and then realized the illusion. Perhaps in a moment of light-heartedness he threw a stone, or his spear, at the zoömorphic figure. Neither flight nor fight being forthcoming, the figure became a target for play – or for practice. The hunter had absolute power over the unmoving target; he could strike it at will. In symbolizing the prey animal, he reinforced his power and mastery over the living animal he would hunt tomorrow.

In one Neanderthal cave, deep in an Italian mountain almost 500 yards from its opening, there was discovered a "vaguely zoömorphic" stalagmite at which, it is thought, groups of Neanderthals cast pellets of clay in some ritual which is presumed to have been magical. The cave was littered with the bones of cave bears. Perhaps the stalagmite resembled the living bear, which was then "fired at" in a ceremonious way, either in recollection of past successes or in preparation for new confrontations.

This type of magic is relatively simplistic. So far as we can guess, it did not involve the intervention in the man/bear relationship of any power other than man's own physical and mental capabilities.

The power of the supernatural came by the time of Cro-Magnon man. The great cave-artist was probably the first practitioner of true magic – the deliberate manipulating of natural objects toward the goals of human will.

Many of the beautiful cave paintings are located in remote recesses. They must have been a challenge to even the most agile of cave men. It would seem that for some reason the artist felt it necessary to retire to these deepest cracks, crannies and crevices to perform his ritual. He retired far from the light of day, from the sight of the animals he portrayed, and from his uninitiated contemporaries.

Many of the finest murals consist of paintings superimposed on one another. In some cases there have been as many as four layers. Why obscure one painting by another? It appears that the magic occurred in the conception of the painting, and each new hunt called for a new one. The places in which superimpositions appear were those which made the best magic.

As the spear is the extension of the arm, so the cave drawing is the extension of the spear. Many of the drawings show animals wounded, with spears protruding from their bodies. Others show animals, such as the mammoth, caught in pits or other traps.

Each graphic representation was an attempt to increase the odds in favor of the hunter. Man and other animals had thus entered into a totally new relationship. Man had ceased to be an integrated part of the natural ecosystem. He now had something which was even more important than flints and spears. In its significance, it ranks only with fire. René Dubos has remarked that the new magic was "probably more important for the understanding of man than physiological and biological knowledge of bodily structure – indeed, more important than the development of tools." From here on, it was the modern era of human dominance.

Whether conscious "propitiation" of wild animals was in evidence by this time is difficult to say. It is well known that men of many societies, before taking an animal's life, have sought its forgiveness in an attempt to ensure that the victim would succumb gracefully, would not injure the hunter and, perhaps more important, would not return to haunt him. Such notions would have to be predicated on the concept of "soul."

Cave art did not stop with hunting magic. The cave people knew that the animals they hunted had to conceive and bear young in order to keep the hunting grounds stocked. Emphasis was given to the fertility of game species. Adults and young were frequently shown. Also, there are illustrations of animals with magnified genitalia and udders, and there are those which are said to suggest the copulation of animals. Cro-Magnon seems to have known the connection between the sex act and birth. He wanted game species to continue to bear offspring, so he used magic not only to secure game but also to ensure the fecundity of game – a Stone Age version of the "sustained yield" concept which only returned to Western human thought after many thousands of years – in fact, in the twentieth century after Christ.

Fertility magic as practiced by Cro-Magnon does not appear to have been restricted to game. Some remarkable figurines have been discovered. These are female statuettes with the secondary sex characteristics – breasts, buttocks, belly – swollen and exaggerated. Apparently Cro-Magnon recognized the need for human fertility as he did for that of the prey he subsisted upon.

Strangely, representations of the human figure are almost entirely restricted to these carvings; they rarely appear in cave paintings. When they do turn up on cave walls, they are usually in the form of relief carvings. Howell remarks, "This is natural, for anyone accustomed to practicing sympathetic magic by making pictures would not be likely to run the mortal risk of having it practiced on him by drawing a picture of himself." Perhaps cave walls were bad magic for people in general. Two dimensions were good enough for animals, but people had to be represented in three! There, it is possible, may be that extra dimension separating man from the rest of nature – flat Cro-Magnon paintings of wild animals, with people in the round.

Although literal or representational pictures of human beings are mostly limited to small sculptures and carvings in relief, there were many cave drawings of *partially* human figures. In these, parts of people were conjoined with parts of animals in a way roughly similar to more recent drawings of mermaids and centaurs. Mermaids and centaurs have human heads and torsos, but Cro-Magnon's animal-men often had humanoid bodies with animal heads and other appendages, Minotaur-like. The Minotaur of Crete was the monstrous get of a supernatural (or, at the very least, superhuman) woman by a bull. But there is no basis on which to impute such decadent fancies to the Cro-Magnon people who, far from being jaded, were very likely much too busy subsisting.

Some believe these animal-men to be the first "gods" of record – the first supernatural beings to have taken literal form in the human imagination. Others suggest they are illustrations of medicine-men summoning up good luck for the hunt tomorrow by putting on the skins and antlers of earlier victims. Perhaps they bounded about in imitation of the living animal's movements. This would be a kind of medicine dance.

These drawings of creatures part human, part animal, may also represent our first excursions into anthropomorphism – the attribution of uniquely human physical characteristics, as well as motives, feelings, and responses, to non-human creatures. Anthropomorphism is a tradition which has been dear to man from Cro-Magnon to Walt Disney. It is the result of man's unwavering determination to interpret all things through himself. Nothing can have any meaning unless it is relatable to man; nothing can have any value unless it is possessed of human qualities. If this is true, then Cro-Magnon was very much the modern man. It must be said, however, that he clearly had an awe of and respect for living, sensate, nonhuman life which was largely forgotten in ensuing millennia.

Hunting societies, both ancient and contemporary, have in common their intimate and inseparable harmony with the natural world around them. Dependent as they are on the mammals, birds, fishes, and other components of the landscape, the people are very much aware of seasonal fluctuations in the abundance of certain kinds of food, and thus of the regular rhythms of the year. Hunters are the first naturalists and ecologists. Their lives depend on their knowing and anticipating the cycles of the seasons and the corresponding requirements of the animals they themselves depend upon. As Rolf Edberg says, nature determines the rhythm not only of the hunter's day but also of his years, "filling him with impressions, creating the framework of his emotional experiences." It would be difficult to categorize the emotional experiences of early *Homo sapiens,* but we can be certain that they were compounded of varying degrees of fear, frustration, anxiety, lust, gut satisfaction – and hope. René Dubos has called man "the hoping animal"; that is as good a definition as any.

If early man had hope, his hope was modest and it was

usually fulfilled. Even in the most ancient times, man had become accustomed to the turn of the seasons and the regular ebb and flow of populations of animals that were important to him. He learned when the fish would run, when the bears would emerge from hibernation, when the berries would ripen, when the caribou or antelope would come – and go. One would suspect that Cro-Magnon had no special need to spend his hopes on events of this kind; they were predictable. More likely, his hopes were centered on the very short-term – on the daily chase. This was where the magic came in. There is every reason to judge that the original magic, low-intensity though it was, worked well enough. There was plenty of game, and our ancestor had become quite an efficient hunter.

But then there would be the unexpected. A member of the band might be killed in the kind of exploding turmoil that must have been mammoth-hunting. Or a child might be carried off by a hyena. No doubt the hunters were accustomed to such misadventures, or at least inured to them. There were other happenings, however, which could not be anticipated – a difficult winter, or a protracted drought. Accustomed as they were to the more or less regular, cyclic progress of nature, men would be baffled, confused, and frightened by a sudden change in the accepted rhythm of things. One might expect also that some of the natural catastrophes, such as floods, avalanches, and the like, would find their way into the spoken history of the tribe, and thus into its culture. No doubt they would be regarded as "unnatural," perhaps even as manifestations of anti-human vengeance or malevolence.

This was the point at which a far more important kind of propitiation became necessary. It was the time when nature came to be regarded as a force which had to be dealt with, for the good of man. In his formidable *The Golden Bough*, J. G. Frazer tells us that "with the growth of his knowledge, man learns to realize more clearly the vastness of nature and his own littleness and feebleness in presence of it." Recognizing the enormity of the forces, Cro-Magnon's art was only a first step in combating nature.

Eventually the cave artist and sculptor evolved into the full-time shaman. In working his spells on cave walls and ceilings he had demonstrated his "power over nature," and that he could not only draw pictures of animals but of animals in human form. His magic had worked well, but now something much more potent had to be devised.

Frazer says, "In time . . . (man) discovers there are things he cannot do, pains which even the most potent magician is impotent to avoid." Magic must evolve into something better, because these impersonal forces conjured up by the shaman "act automatically and obey whoever applies the right formula." Impersonal forces which could not be properly identified, much less controlled at will, were not sufficient to deal with the larger issues of life. This was the point at which magic began to give way to more personal, homocentric forces aided and abetted by supernatural beings. This was the dawn of religion, and of the evolution from sorcerer to priest.

The role of the priest was to influence the often capricious will of these supernatural beings, by various rituals, familiar incantations, sacrifices, and prayers. The painted anthropomorphic figures on Cro-Magnon cave walls indicate the transition of the painter from magician to sorcerer to priest.

With the advent of religion, man was further alienated from nature. The more human these deities became, the wider the gulf between man and nature. By propitiating and influencing the gods, man was imposing his will upon surrounding nature and, by extension, upon the cosmos.

Man has spawned an infinite number of religions and extra-human fancies, but in the matter of the contribution of religion to man's indifference to nature, none is more interesting than animism. It is customary to link animism with the pejorative "pagan." Pagan animism was all about souls.

There must be countless definitions of "soul," but if we accept the most fundamental meaning of the word to be the life spark, then every individual human being, nonhuman animal being, and plant being, has a soul. That is incontrovertible. Apparently, Neanderthal man believed it. With his burial ceremonies he recognized the termination of a unique being, the departure of life, the departure of a "soul." He came to realize also that his own life spark would one day leave him. Perhaps even in Neanderthal times man recognized that other things – caribou, bears, blueberries – had souls too.

One may describe this type of animism, encompassing all forms of life, as the reverence preached by Schweitzer. Animism and Schweitzer's reverence for life are one and the same. Yet the kind doctor, a devout Protestant Christian, could never have allied himself with pagan notions. Indeed, according to his religion, no nonhuman creature could possibly have a soul. There have been many ironies.

Frazer emphasizes a significant transition in the evolution of religious thought. "When a tree comes to be viewed no longer as the body of the tree-spirit, but simply as its abode which it can quit at pleasure . . . animism is passing into polytheism." He explains that instead of seeing the tree as a sensate, living being, the man now regards it as "a lifeless, inert mass, tenanted for a longer or shorter time by a supernatural being who, as he can pass freely from tree to tree, thereby enjoys a certain right of possession or lordship over the trees, and, ceasing to be a tree-soul, becomes a forest god."

This is significant in attempting to pin-point the "fall of man." The tree or the deer is no longer a free entity like the man. Man and nature have become entirely different things. The next step, the anthropomorphic one, was to vest the forest god with the body of a man. Man's power over nature was given an irreversible rocket thrust. The loss of animism and the substitution of theism was one of the most critical turning points in history.

It is likely that the domestication of animals and plants was a substantial contributing factor to the elaboration of various supernatural rituals. Fertility of game had been important to man the hunter; fertility of domestic flocks and plantations was of equal importance to man the pastoralist and agriculturist. The earliest dates of domestication are vague, but it seems to have occurred just at the end of the last glaciation some 10 or 11 millennia ago – during or immediately after the peak period of Pleistocene overkill.

We have long fancied that the dog must have been the first domesticated animal, and that man's best friend has been around for as long as we have been capable of interspecific friendship. We have made many speculations about the little motherless wolf puppy brought back to the cave, there to be showered with affection, rescued from the world of savagery and reared in the warm benignity and togetherness of man's world. Man the hunter and dog the hunter forged an inseparable association and relationship to their mutual advantage. The pleasant symbiotic bond has persisted to this day.

There is little evidence for this. Contemporary hunting societies do not use dogs for the chase (Eskimos will loosen their sledge dogs at the critical moment, to bring a bear to bay, but they do not *hunt* with dogs), and there is no reason to believe that early hunting societies did so. If anything, man and wolf are competitors. The likelihood is that they seldom even encountered each other. On those occasions when they did meet, chances are that the wolves behaved much as they do today – they vanished in an instant. Wild dogs do not like human company, and there is no reason to assume that they ever did. Dogs are intelligent, and like big cats and other large mammals, they learned long ago to stay out of man's way.

The more reasonable assumption is that the dog was first domesticated as a source of food. Dogs are still eaten in some societies. There is no "instinctive" man/dog bond. One need only notice how dogs are treated in most parts of the world today, both in non-industrialized and in "developed" societies. I have seen injured and helpless dogs teased, abused and tormented by laughing children in an African desert and in the heart of one of South America's greatest cities. One can imagine that dogs and other domestic animals received no better treatment at the hands of our antecedents.

The sheep appears to have been the first domestic animal.

There is a record of a recognizably domestic sheep from 11,000 years ago at Zawi Chemi in Iraq, a mountain site in the east drainage of the Tigris River. Where there were sheep, no doubt there were also goats. Swine are also very old in the history of domestication. These are all quite gentle creatures, and logically would have been the best subjects for domestication. One might guess that the husbandry of these animals, and of dogs, led to the knowledge and ability to deal with large, dangerous beasts such as cattle.

It is possible that cattle were domesticated initially not for meat or even for milk, but as symbols of status. Cattle have a long history in this regard, which persists today in such areas as Masailand in East Africa. A large, ponderous and fierce-looking animal such as a wild bull (or even a cow) would be a conspicuous advertisement of the authority, strength, bravery and thus prestige of the man who owned one. There is also a possibility that even in the earliest times cattle may have had a religious rather than a utilitarian or economic significance.

The domestication of animals was another large feather in the growing warbonnet of the human species. Power over nature had long been associated with magic ritual and the killing of wild prey. The new ability to maintain herds of animals under one's immediate control, subject to one's slightest whim, must have been an enormous boost to the collective ego. The rewards of supremacy could never have been more satisfying than when it was learned that certain animals could be taught to obey human commands.

It has long been thought that the domestication of animals and plants, which we now term the agricultural revolution, began in a relatively confined area in the foothills surrounding the great valley of the Euphrates and the Tigris, and that it spread outward from that nucleus. Recent discoveries indicate that domestication probably did not arise in any one place and that it was not accomplished by any single group of people. According to Chester S. Chard, "Probably each crop and animal developed in a different part of the region and then eventually coalesced to form the standard farming complex when enough alteration had taken place genetically so that each crop could spread beyond its original habitat niche." The first plants to be "tamed" could only be cultivated in the area to which they were native, and it must have been a great many years until artificial selection, whether it was deliberately or unconsciously applied, had produced varieties which could be transplanted into new ecological situations.

With the beginning of agriculture, as with the domestication of animals, the relationship between man and nature entered a new phase. The uncertainty of the chase had been replaced by the certainty of reproduction. The laws of nature were being used in the propagation of captive plants and animals, to be sure, but their very use and the confinement of a variety of living things led to a new type of human social behavior and a new attitude toward nature as a whole.

By comparison with the time when he had been a hunter and gatherer, man was increasingly independent of external forces (although he soon learned about continuing requirements like soil and water). This had its cultural significance. As J. D. Bernal has described it, the "new kind of society was qualitatively different because of the quantitative increase in the number of people the same land could support." He adds that the transition from hunting to agriculture was the transition we now know in legend as the "fall of man"! Man left Paradise or Eden (the happy hunting ground) to take up working for his bread by the sweat of his brow.

Agriculture meant permanent settlements, which in turn led to what has been called the "institutionalization" of religion in both theory and practice. Power over the animals had become power over the landscape; and since one was now dealing with natural forces in a broader and continuing sense, religious rites became increasingly important in the community. Much attention must have been given to fertility – of animals, of plants, and of people. Religious rites were largely devoted to sex and reproduction, and human sex acts were

encouragements for the crops to do likewise, and to be boun-
tiful. The concept of overcoming nature by sheer pressure of
numbers would seem to have begun very early; it is no wonder
that even in the twentieth century it has been difficult to
overcome ancient taboos in connection with the limitation of
human numbers.

In the first days of agriculture, embryonic communities
were dependent on seasonal rains and flooding. This gave a
cyclic, predictable nature to their lives, and was a prominent
part of the religions of the time. Prayers and other rituals
devoted to rain were not different from the earlier forms of
magic, which were still widely practiced in spite of the emerging organized religions. But magic was being rapidly down-
graded. Frazer says magic "is now regarded as an encroach-
ment, at once vain and impious, on the domain of the gods,
and as such encounters the steady opposition of the priests,
whose reputation and influence rise or fall with those of their
gods."

It was possible, however, to help the gods. Even then it
was known that the gods mostly help those who help themselves. Artificial supplying of water – irrigation – came quite
early, and it united numbers of nuclear communities into
cooperative ventures. It was a different kind of agriculture
which emerged. A number of villages working on one irrigation project were drawn together into a supervillage, and it
was a short step from the supervillage to the city-state. More
and more people could live on the land, and the greater
number of people meant a new kind of social organization.

It seems likely, on the basis of evidence which has been
unearthed in even the earliest communities, that "religion
functioned as the original political catalyst in the formation of
urban societies and that the early states were strongly theo-
cratic in character, with the priesthood in fact forming the
ruling class" (Chard). The progress from priest to king to
emperor is well known. And power over men was also power
over nature.

CHAPTER TEN

The Ladder of Perfection

The conceptual separation between man and nature goes back to the development of tools and weapons, fire and social predation long before the appearance of genus *Homo*. It involves early human magic and cave drawings. Man considered himself apart from nature long before he could consciously articulate the idea, much less before it could be codified, dogmatized, and institutionalized.

Some writers have singled out Christianity as the key factor in modern homocentricity and the origin of all that is worst in our attitudes toward our living environment. It is not as cut-and-dried as that. Christianity, and Judaism, and all other religions and schools of philosophy were vectors or channelers of age-old assumptions – vehicles for the cultural transmission of traditional ideas that have rarely been questioned since the origins of human thought. Nothing biological or cultural remains static. All things evolve and are modified as they evolve, and the churches and schools of thought undoubtedly modified our attitudes. But the germs of the evolution of modern homocentric thought long antedate the Torah, Christianity, and the medieval scholars.

For as long as human societies have existed, people in those societies have attempted to identify a relationship between themselves and the observable facts of the cosmos – the cycles of the seasons and the years, the apparent movement of the sun and the stars, and all the other natural phenomena surrounding them. It was vitally important to explain these things. The mere fact of posing the question, of course, sets man apart from that which he is questioning – the nature he is attempting to relate to. In fact, most such inquiries have been made from the original bias that there are three components to the universe – man, nature, and the supernatural. This fundamental block to productive inquiry has plagued us from the beginning.

Some of the more fanciful cosmogonies are quite attractive, and in moments of light-heartedness one can approve of the healthy skepticism of the Flat Earth Society, for example. One rather likes the old Hindu notion of the giant tortoise on which stand elephants which in turn support the earth. Or the Egyptian picture of the universal egg. One of the Phoenician theories involved spontaneous emanations giving rise to organic matter, to which one can only say, "Why not?"

There is a Polynesian myth of a god of the air who hovers over the waters of the world, breathing life into things. From there it is not a great jump to the Zoroastrian view of a personal deity exercising his own free will in the creation of things; and to the ancient Babylonian creation story, which is so strikingly similar to the opening passages of Genesis. Plato himself conceived of a personal creator. It is the evolution of these anthropomorphic forms and anthropocentric notions to which we must address ourselves in order to comprehend the maturation of man's self-appointed lordship over the planet.

Some of the ancients did in fact conceive of human responsibility in environmental questions. As all conservationists know, Plato bewailed the disappearance of the forests from the slopes of Greek mountains, and noticed that as the result the rainwater ran off directly to the sea without making its contribution to the soil and thus to domestic plants and animals. Clarence J. Glacken in *Man's Role in Changing the Face of the Earth* remarks, "Perhaps if this view had found more elaborate expression in the *Laws,* in which Plato discusses the origin and development of society, an awareness of the philosophical implications of man's activities in changing the environment might have at that time entered the mainstream of Western thought."

Unfortunately, it did not. Mostly, perhaps, because the poet Plato was essentially an idealist and an academic rather than an observer of nature, and although he was clearly aware of the effects of deforestation he was much more interested in his philosophic pursuits. Equally or perhaps even more unfortunate was the failure of the great natural scientist Pythagoras to penetrate and influence the trend of human ideas.

Pythagoras, whose "life-style" was the objective investigation of nature (and who, my colleague C. B. Cragg says, "inadvertently invented Western civilization"), knew that in natural rhythms and relationships there are the "harmonies that all men must seek." As it turned out, Platonism was much too potent for the Pythagorean view of things, and, as Bernal says, conformist Platonism "held back knowledge of the real motion of the heavens, and with it any possibility of valid physics, for 2,000 years." And, of course, man was considered the highest product of creation and it was felt that everything had been created expressly for his purposes. These assumptions were not unique to Greek culture; they can be traced back to some of the earliest Egyptian myths of creation.

Man, the ultimate and highest life form, is unequivocally described by Aristotle. Aristotle's model of the world of nature did not involve either creation or evolution, but it consisted of the familiar scale, starting at the bottom with minerals and soils, then vegetation, then more and more "perfect" animals, with man at the summit. Aristotle took it for granted that there can be no purpose for anything in the world except for human deeds and human artifacts. This notion echoes and re-echoes through human thought, even into the twentieth century, reinforced as it has been by successive dogmas through more than two millennia. It must be said, however, that the generations of churchmen who advanced Aristotle's homocentricity managed to overlook one fact: when he announced that there was no purpose for anything save man, he meant it. There was no purpose for God, either.

In his fascinating *Aristotle,* John Herman Randall, Jr. points out, "The one thing the mature Aristotle did not understand, and apparently had no interest in investigating, was religion. This makes the use of his thought by the great medieval traditions as a religious apologetic seem a colossal irony." Aristotle invested no time in wondering how the world was made, how it evolved, or whether it was created. It was perfectly reasonable in its present form. Man was man, all was there for his examination and comprehension, and that was all that mattered. There was no need for it to have been created.

There was another side to Aristotle. That was his curiosity, and his remarkable insight into the nature of things. This side has been largely forgotten. There is ample evidence that in certain contexts he was uncannily close to ideas which in the twentieth century are considered frontiers of human thought.

Aristotle was essentially a man of science. He recorded things he observed, and he was not concerned with how they got to be that way. Since he did not recognize evolutionary processes, he saw species as immutable; species were more or less possessed of "perfection" according to the various strata in which they found themselves. Aristotle was fascinated, however, by the ways in which different kinds of living things seem to be *fitted* to their distinctive life patterns. This was his observation of the adaptations of animals to their environments; but he saw these adaptations as permanent, fixed, and unchanging.

The Aristotelian world was one with no beginning and no end. Seasons come and go, and repeat themselves; animals are born and in turn give birth, and die, in an unceasing cyclic pattern. The same applied to the cosmos, which was "single and eternal, having no beginning and no end of its whole existence, containing and embracing in itself infinite time."

Significant about Aristotle, however, was that his anthropocentricity was greatly tempered by humility. Randall says, "Aristotle's whole life, as the scholars have now reconstructed it for us, was the realization that intellectual imagination, having plenty of brilliant ideas, is not enough. We need sense also, humility in the face of encountered facts." His unshakable belief – that we must observe, and deal with observed facts, with what actually *is* – endeared him to men such as the deeply religious Charles Darwin, who commented, "Linnaeus and Cuvier have been my two gods; but they were mere schoolboys compared to old Aristotle."

Surely it is one of the intellectual tragedies of all time that

Aristotle, in his investigations so incredibly close to a man/ nature ethic for his time and ours, was corrupted and perverted by ensuing generations of scholars. Careers have been made of disputing Aristotle. Bernal claims that after Bruno and Galileo, the history of science was largely a story of how Aristotle was overthrown in one field after another. He spoke too soon, and the famous sixteenth-century Frenchman Peter Ramus spoke *much* too soon with his sweeping remark that "everything Aristotle taught was false."

Aristotle knew that the study of observed facts was the proper role for mankind, and that it is pointless to search for mystical origins, supernatural beginnings and ends, when the eternal cycles of nature are everywhere to be seen and comprehended. Irving Babbitt observed that if the champions of the modern spirit who rejected Aristotle "had been more modern they might have seen in him rather a chief precursor. They might have learned from him how to have standards and at the same time not to be immured in dogma."

It is sad that the best of Aristotle was lost and the worst perpetuated. But there is a long human history of original ideas being changed in translation, with the orthodoxy of one age becoming the heresy of the next. By the time of Cicero, man was immovably planted at the center of the universe. "We are the absolute masters of what the earth produces" by virtue of our position at the summit of nature's ascending scale. Cicero was the thoroughgoing technocrat, and he would have felt very much at home today: "We stop, direct, and turn the rivers: in short, by our hands we endeavor, by our various operations in this world, to make, as it were, another Nature."

Aristotelian worship of the human intellect had its extension in the philosophy of the Stoics, and at one period it seems that there might yet have been hope for a rational philosophy of nature, a spiritual man/nature relationship of the kind that Schweitzer much later proposed. Unfortunately, Stoicism managed to emphasize and underscore the gap between man and the rest of nature by its insistence that man alone has a rational, intelligent soul, and that animals act only by instinct. Intellectual rationalization of man's oneness with the cosmos did not materialize.

Had the spirit of free inquiry into natural phenomena advocated and practiced by Pythagoras and Aristotle been permitted to continue, it would have soon been relatively apparent that man was not in fact at the center and was a functioning and integrated part of the natural system that was beginning to be understood.

Even more important was the cyclic philosophy of ancient times. Had this been allowed to persist, with its deep and abiding faith in the repeating and immutable nature of things, Judaeo-Christian (and Baconian-Cartesian) "progress" might conceivably have been avoided. The beauty of Greek thinking was its recognition of natural perfection. Things *are,* and they are beautiful. The more we look into and understand the natural sciences, the more we may understand about man. To our misfortune, the monotheistic myths of the Torah, borrowed from ancient Sumeria and Babylon, complete with beginnings and ends, overwhelmed the earlier culture. The new "nonrepetitive and linear" concept of time, completely foreign to the Greek and Roman way of thinking, was to inherit the Western world.

The Torah was said to have been handed down to Moses on Mount Sinai as the Law. The earliest date for the Torah is 444 B.C., two generations before the birth of Aristotle; and there had been a long Hebrew tradition prior to that time. The Book of Genesis is said to deal with the period between 4004 and 1635 B.C., and the injunctions contained therein had long been an integral part of Judaean culture: "Be fruitful, and multiply, and replenish the earth, and subdue it: and have dominion over the fish of the sea, and over the fowl of the air, and over every living thing that moveth upon the earth."

Genesis is a reflection of deep-rooted human convictions, and a component of the basic assumptions that were to be developed in both Jewish and Christian traditions, and thus in Western thought. It postulated a personal God, as Plato had done; it placed man on the uppermost rung of the ladder of living things, as Aristotle had done; it gave the final and ultimate authoritative sanction to man in his role as master of the world.

In his famous essay *The Historical Roots of Our Ecologic Crisis,* Lynn White, Jr. writes: "God planned all of this explicitly for man's benefit and rule; no item in the physical creation had any purpose save to serve man's purposes." It was an ideology easy to accept, for it answered questions that had pestered the best human minds since the beginning: "Who am I? Why am I here? What is my place?" Man, in the image of a personal and all-powerful and benevolent deity, was given mastery of all that was nonhuman. Man had an immortal soul. Non-man had no immortal soul. All things were caused and created to benefit the earthly image of an anthropomorphic God. To this was added Christianity.

The first chapter of Genesis is not the whole of it, nor is Genesis the only book in the Bible. As every scholar knows, and as almost every schoolboy knows, there is a potentially infinite number of interpretations which can be made of both Testaments. The "dominion/subjugation" roles of man and nature as spelled out in Genesis are counterbalanced elsewhere by the notion of "stewardship." Many academics and churchmen have called attention to this, and continue to do so. One of the earlier and thus more noteworthy references to this theme is that of Sir Matthew Hale in the seventeenth century who, while acknowledging that man was at the summit of the (Aristotelian) scale, still felt that he had concomitant responsibilities, for man is a caretaker, a "steward of God." It probably need not be added that this was an aberrant posture for his time.

The idea of "stewardship" is this: God has given man the planet Earth and everything on and in it, but God has also given man the responsibility to keep all these things in as good condition as they were when he received them. Man is God's steward. This is undeniably present in the Bible, but it is no whit different in its anthropocentricity than any of the other "God, man, and nature" references and allegories. Also – and this is even more important in practical terms – it has never been *acted upon*.

Throughout Christian history, the style of Western man's dominion over nature has been that of Genesis 1, and the few sensitive and hopeful souls who have properly recognized the relatively minor theme of stewardship have had no impact on public affairs whatever. The world is full of Christian apologists who call attention to the stewardship theme in defense of their persuasion, but their recognition of it has never progressed beyond the academic stage. It has made no impression upon the "real world." What *has* made an impression is the sense of Genesis 9:2 – "and the fear of thee and the dread of thee shall be upon every living thing. . . . "

Orthodox Christianity is a long, long way from the living world which gives us breath and substance. It revolves first about the existence of one God. The fact that man is made in God's image indicates that this god, like so many of his predecessors, is an anthropomorphic God. He makes it that much more difficult for us to remember man's kinship with nature. Man and God are able to have their private discourse, from which nature is excluded.

Secondly, there is the concept of "original sin," and the "fall of man." This is tantamount to the "killer ape" theory. Our origins are rooted in blood crimes of such appalling enormity that none of us is fit for the "service and fellowship" of God; we are subject to God's displeasure now and in infinite eternity. That there is in fact no "good" or "bad" in nature is ignored or suppressed. If you want to keep them in line, you must keep them frightened.

Finally, since man is so inept, and is incapable of extricating himself from his heritage of wrongdoing and suffering, God sent his son "to save sinners, to deliver them from hell, to make them holy, and partakers of the eternal joy and glory of heaven."

The aggregate of Christian thought is even more man-

centered than the Hebrew faith which preceded it. There is
no room, anywhere, for man, the extraordinary product of
primate evolution. There is no room for man, the glorious
artifact of living nature. All is suppressed to the categorized
picture of the world, with the whole of nature having been
created by God separately from man and for man's (and
God's) glory. Whether man is regarded as total and absolute
master and consumer, or as steward, does not really matter.
The dichotomy is clear. The descending order is God-man-
nature. There is no conceivable reason for the existence of
the blue planet apart from the needs of God and man.

One of the most elaborate and complex rationalizations of
Christianity, nature, and man was presented by Pierre Teilhard de Chardin, who died in 1955. A French Jesuit, he was
also a geologist and a paleoanthropologist of distinction, having been involved in the early discoveries and interpretations
of *Sinanthropus,* now *Homo erectus.* One's impression of Teilhard is of a brilliant man of science and religious faith in equal
measure, and thus a man who was intellectually tortured. He
was a philosopher who bent every energy of his immense
talent to making "sense" of mankind in the light of twentieth-
century scientific knowledge, yet in the sense, also, of the
dogma to which he was committed. Perhaps one could say he
was dedicated to the reconciliation of faith in man and God
with faith in nature and demonstrated natural processes. He
was also a mystic. The turns and ramifications of his reasoning
are not easy to follow, especially since his style is enormously
complicated and he found it necessary to coin words to convey his meanings. But he is worth the effort.

Teilhard recognized the fact of evolution. As a Christian,
however, he saw man as "the only absolute parameter of
evolution." Evolution existed solely to produce mankind. All
of Teilhard's immense intellect was directed, it would seem,
toward constructing an anthropogenesis, an explanation for
the phenomenon of man. (Having in mind his background, we
must accept that an explanation for man *had* to be found; the
majestic accident of chance evolutionary process was not
enough.)

For Teilhard, as for Aristotle, man was "the unique, terminal, inflorescence on the tree of life." Man's evolutionary development was purposeful; it was toward a defined and absolute goal, "polarized towards a determined point." All of evolution, before the appearance of man, was thus determined; the evolution of man, in its ascending progress, raised the human mind "to the position of constituting a specifically new envelope to the earth." This new envelope, superimposed on the biosphere, was the envelope of *knowing* – the *nöosphere.*

In Teilhard's view, a consciousness pervades everything. In the course of evolution, this consciousness has become more and more highly organized as it has moved up the scale of vegetative and then animal complexity. Always, as with Aristotle, it has moved toward greater and greater perfection. In man, we have the "ultimate and supreme product" of the first stage of evolution.

Man, according to Teilhard, is alone among the animal kingdom in his ability to be conscious of himself, or to reflect. This self-consciousness of man represents the envelope of knowing – the nöosphere.

Teilhard sees the next stage of evolution as the achievement of "collective reflection," or universal knowing – a complete and unified consciousness of all mankind. The goal of this determinist evolutionary progress is an "Omega point." The design is God's, and the achievement of universal consciousness is man's function in it. Teilhard expects that the attainment of collective reflection may require "some millions of years." Ideally and theoretically, "mankind will come to an end when, having finally *understood,* it has, in a total and final reflection, reduced in it everything to a common idea and a common passion."

This is a remarkable statement from a man of science in the twentieth century. It is a demonstration of the infinite capacity of a rarely gifted mind to perform intellectual contortions within the confines of traditional and indisputable dogma. But it is impossible to ignore Teilhard's thesis. Every one of us at times raises the questions, but rarely do we attempt to answer them with the persistence and the elegance – much less the faith – of a Teilhard. His message regarding world-wide understanding is abundantly clear.

Yet, one is obliged to recognize the frightful assumptions which underlie Teilhard's assessment of man's role on Earth. All was placed in readiness by divine plan before the evolutionary "progress" was set in motion, and the master design could only be realized in the form and mind of man. Teilhard says, "Man is irreplaceable. Therefore, however improbable it might seem, he must reach the goal, not necessarily, doubtless, but infallibly." Since there is no reason for the existence of anything on Earth save for the human mind, all else is inconsequential.

We have been entertaining such thoughts, and expressing them, as long as we have had thoughts and the power to express them. The man-centered universe is almost as old as man himself. Our traditions have done their work so well that, as Teilhard discovered, we do not even have the vocabulary to articulate new ideas, much less communicate them.

One would wish that our species had been granted somewhat more humility and somewhat less consciousness of self, or "reflection." Or perhaps one might also wish for more open-mindedness and a great deal less of the capacity for self-deception.

CHAPTER ELEVEN

Zeal for Subdual

The first fifteen centuries of Christianity were a dark and empty void in human awareness of nature. Evolution of objective thought and honest reflection were largely choked off by the medieval church.

The literate minority of people was mostly preoccupied with dogmatic interpretations, often erroneous, of Aristotle and other classical scholars. Scientific and even academic excursions into the living world were not set in motion until the latter part of the period, the dawn of the age of exploration and conquest – both, as it turned out, chiefly in the name of the church. In its final decades, however, the sixteenth century became a kind of intellectual crucible out of which poured the devastating thoughts of Copernicus and Bruno, Galileo and Kepler, who, with the unparalleled Shakespeare, led men's minds into the first fresh air of modern thought.

There is an astonishing gap in the development of natural science between the time of Aristotle and that of Descartes and Bacon. No one was able, or permitted, to pick up where the classical scientists and philosophers had left off. Well before the time of the birth of Christ, philosophy had become something quite different from science and politics, and had taken on an almost entirely moral or ethical tone. Where the philosophers of classical times had been fascinated by the world of nature and the intriguing questions in the observable cosmos, the philosophers of the Dark Ages had become priests; and "science," if it was such, represented what Bernal says might be called the religion of the cultivated upper classes.

The stultifying effect of the medieval churchmen on the growth of human knowledge is epitomized in the fate of astronomy, the most ancient of the sciences. Long before the Christian era, astronomy was known to Chinese, Hindu, Chaldean, Egyptian and Greek philosopher/scientists. Its interpretations varied; all except the Greeks seem to have used their observations of the heavens in the pursuit of the pseudo-scientific aims of astrology. The Chinese made political extrapolations from their observations, while the Hindus, Chaldeans and Egyptians found various ways of building astronomy into their religious structures.

The Greeks made astronomy the first of the exact sciences. Six hundred and forty years before the birth of Christ, Thales, building on the work of the Egyptians, was able to predict a total eclipse of the sun, and to call attention to the value of the constellation *Ursa Minor* (the "Lesser Bear") as an aid to navigation. A practitioner of abstract geometry, Thales could calculate the heights of things and the distances of ships at sea. He was also sufficiently free-thinking to believe that water was the element from which all things had evolved. From his time, through five centuries until the days of the great Hipparchus, who died in 120 B.C., the ancients made extraordinary advances in science, which at about that point were abruptly terminated for fifteen hundred years.

Shortly after Thales, Pythagoras had promulgated the notion that the earth and other planets move around the sun, but he was not to be recognized for many centuries. It has been said that if these ancient scientists – especially Hipparchus – had had access to the telescope and the pendulum, it would have taken no more than fifty years to bring astronomy to the point at which it was at the time of the birth of Isaac Newton in 1642. Apart from Ptolemy in the second century of Christ, there was no astronomer, and no advance in astronomy, until Copernicus, who died in 1543. He had picked up the motion of the planets where Pythagoras had left it two thousand years before.

It is clear that the Dark Ages were a total loss to the development of human comprehension of the world men live in. The churchmen had kept man firmly and immovably locked at the center of the universe. The homocentric Christian view of the world was so all-pervasive in Europe, and so unquestioned, that even the few men of science who emerged from time to time during the largely barren centuries were themselves unable to conceive of any structure of the universe other than the Aristotelian ladder of lesser-to-greater perfection, with man securely perched on the topmost rung. And no matter where a man found himself in the hierarchy of society, from peasant to priest, he had no doubt whatever of his personal dominion over nonhuman nature.

Medieval thought was dominated in the thirteenth century by St. Thomas Aquinas, the Italian scholastic philosopher. Aquinas, maintaining that theology and science cannot be mutually exclusive or contradict each other simply because truth is indivisible, set about the enormous task of reconciling the Aristotelian heritage with Christian dogma. As we have seen, the purest of Aristotelian thought does not allow for beginnings and ends, much less for personal Creators, still less for religion of any kind. But the brilliant synthesis of Aquinas, like the rationale of Teilhard, made the most of Aristotle's ladder of ascending rank, to which was added God. God, man, and nature were rigidly compartmentalized.

The essence of Aquinas' teachings appears to be that there are two sources of knowledge – Christian faith with its attendant mysteries, and the truths arrived at by human reason. In effect, since truth is indivisible, this means that the mysteries of Christian dogma are to be accepted. They need not be understood to benefit us.

Thus, Aquinas managed to set up a body of dogma which allowed symbolic mysteries and mystical analogies to be accepted as part of the eternal truth. Julian Huxley has commented that this results in the analogies and symbolisms "being taken for more than they are – for scientific knowledge, or even for an absolute certainty of some still higher order – and conclusions then drawn from it." These conclusions, Huxley observes, follow with "full syllogistic majesty," but since the premises are false, the exercise is dangerous.

In his brilliant *Civilisation*, Kenneth Clark states that Western civilization could be argued to have been the creation of the church. He notes that he is not thinking, "for the moment, of the Church as the repository of Christian truth and spiritual experience: I am thinking of her, as the twelfth century thought of her, as a power – Ecclesia – sitting like an empress." Of this there is no doubt. The power of the church in medieval times is almost incomprehensible. And the church did not allow for any ethic that extended beyond its dogma.

Schweitzer has remarked on the "medieval contempt for the world and for life," an attitude brought about by the harmonizing of primitive Christianity and Greek metaphysics. "It is fundamentally a rejection of the world and of life because the interest of Christianity at this time concentrated upon otherworldly things." The church was all-powerful, and the church was entirely occupied with the mysteries of its dogma. Schweitzer added that this early form of Christianity also imposed upon people an attitude toward life that was in contradiction to their nature. They endured it, however, for an exceptionally long time.

We can only guess at the thoughts that occupied the mind of a medieval peasant, and we can only guess at his nature, and whether it was being contradicted. No doubt the peasant believed; he had no alternative. He could not read his own language, much less Latin, so he accepted what he was told of divine revelation. No doubt he also believed in the intercession of the saints on his behalf – or at least he realized it could do no harm to propitiate them. He must have dwelt in constant fear of reprisals, here and in eternity. That this fear transferred itself from hellfire in the future to the living presence of priests in the here and now seems certain. Trevelyan tells us of the typical English peasant of the time of Chaucer, who could not follow the Latin liturgies yet was able to absorb a great deal from his surroundings. "Fear of hell was a most potent force, pitilessly exploited by all preachers and confessors, both to enrich the Church and to call sinners to repentance."

There is little way of knowing how medieval people felt about nonhuman nature, or whether they felt anything about it at all. Since people were the unique possessors of immortal souls, however, no doubt the nonhuman elements of the landscape were viewed with mild curiosity, if not contempt. Perhaps it was indifference. One does not need to suppose that medieval man was actively suspicious of nature, but he certainly did see it as alien.

It is interesting how medieval art abounds with pictures of birds and other animals, and of flowers and trees. One had always fancied that this was merely a manifestation of the artist's enthusiasm for embellishment; but there is more to it. The philosophers of the Middle Ages called it *natura naturans* – nature "doing its thing" – flowering, burgeoning, producing. Not far, really, from "pagan" expressions of nature in fertility rites. But, as Clark points out, "The odd thing about the medieval response to nature was that it saw all those things in isolation. It quite literally couldn't see the wood for the trees." That probably explains one's reaction to them as mere decoration; they have no perceptible meaning. Clark adds, "As for birds, they were a medieval obsession. . . . I think the reason is that they had become symbols of freedom." We still use the expression "free as a bird" even though we now know that birds are among the more narrowly "programmed" of animals, with rather little freedom of individual choice built into their high-intensity careers.

Birds make us recall St. Francis of Assisi who in the early years of the thirteenth century preached to the birds, and is said to have tamed a marauding wolf at Gubbio through gentleness. The wolf repented of his ways, "died in the odor of sanctity, and was buried in consecrated ground."

In his understanding and acceptance of the oneness of all living things, and in his comprehension of the value of both individual and collective humility under God, Francis was a rare medieval Christian. As Lynn White, Jr., puts it, "Francis tried to depose man from his monarchy over creation and set up a democracy of all God's creatures." In this, he was doing essentially the same thing as the modern environmentalist or naturalist is doing – trying to create a perspective, in nature, for man as a part of nature. He was unsuccessful.

To accept this interpretation of Francis' teaching, one

must reject the notion of man's uniqueness in the possession of immortal soul, and extend the concept of soul to all living things. This was impossible within the confines of church dogma, and this portion of St. Francis' tradition has not survived. His vows of poverty, chastity and obedience have of course been perpetuated in the distinguished religious order which bears his name.

Also out of Italy, with its beginnings about a century after Francis of Assisi, came the Renaissance, and the first new stirrings of "humanism," although it reached full flower two hundred years later. The magnificent art and architecture of the Italian Renaissance were unrestrained celebrations of man. Man was equated with beauty, with joy, with dignity – and with power.

Above all else, it was power. The great Venetian architect of the fifteenth century, Leone Battista Alberti, whom Clark has described as the "quintessential early Renaissance man," expressed it clearly. "A man can do all things, if he will." The men who made such utterances must be judged in the context of their time. The Renaissance was not a period for humility. As artists and scholars gradually rediscovered Greece and Rome, and as they became emancipated from the yoke of medieval religious authority, it was natural that they rejoice in their newly discovered intellectual and artistic freedom. Man was liberated, and his mind was free to assimilate whatever touched him, whatever would contribute to the edifice that he was building – the edifice that was to be the monument to man's intellect.

The Renaissance rediscovered classical literature and classical science. Inquiry into the nature of the world, and its processes, could no longer be suppressed. Developments in mathematics and physics had almost instantaneous results in a practical way. No applied science of the time was to enjoy such immediate benefits as navigation.

There was a neo-classical interest in cartography. By the beginning of the sixteenth century, it became possible to work out longitude. Somewhat later, the emergence of such new

hardware as the astrolabe, the sextant, and the quadrant allowed accurate measurement of the position of the sun and
the stars, and thus a reasonable indication of the position of
oneself. By about 1600 came arithmetic, trigonometry, and
the use of logarithms. Thus, to the creative side of the Renaissance was added the extra dimension of sea-faring and exploration. The world was conquerable. No element of the
rebirth of knowledge and the resurgence of man's capabilities
was to be so important as the susceptibility of Earth to conquest.

In the last decade of the fifteenth century, Bartholomew
Diaz had rounded the Cape of Good Hope, Columbus had
reached the Bahamas and parts of the West Indies, and Vasco
da Gama had made it to India. Vincente Pinzon had discovered the prodigious mouth of the Amazon, Amerigo Vespucci
(for whom the new hemisphere would be named) had encountered the Guianas, and John Cabot had reviewed the coastline
from Greenland to Virginia. Within twenty years, Magellan's
party would circumnavigate the globe, Balboa (not stout
Cortes) would cross the isthmus of Panama to the Pacific, and
the implacable *conquistadores* of Hernan Cortes would be sav-
agely destroying the ancient and brilliant civilization of the
Aztecs in the name of church and king. European man had the
entire world in his grasp.

During the great age of exploration, no parts of the world
were more vulnerable to conquest than oceanic islands. The
physical and biological isolation of such islands and the re-
sultant genetic "wanderings" of their living inhabitants had
resulted in some singular evolutionary products. Most important in this regard is the fact that many animals of such islands
have let down their defenses, as it were. They have become
adjusted to a very local and specific set of conditions. They
have become completely dependent on a unique way of life.

Many birds of remote oceanic islands have lost the power
of flight. It is not necessary to be able to fly if a bird has no
land-based enemies. Thus, where there are no land predators, there are the penguins of the Antarctic, the flightless

cormorant of the Galapagos, the various flightless rails of New Zealand, Laysan, Tristan da Cunha, and elsewhere. The North Atlantic greak auk, now extinct, was also flightless.

All of these are descended from birds which once could fly. But when the need for the power of flight was eliminated – in the form of the absence of natural enemies – natural selection disposed of flight. It is said, in fact, that flight may even be disadvantageous to an island-dwelling bird, and the incapacity to fly may have positive survival value. It is quite possible that not being able to fly keeps you down at ground level, and not up in the air where any vagrant storm could pick you up and whisk you away. A number of birds have "chosen" that direction. Others have "traded" the power of flight for specializations in other and even more remarkable directions. The penguin now flies under the water, and the ostrich has become geared to great size, great strength, long legs, and running power.

The most noteworthy of the birds which lost the power of flight were the dodos and solitaires of Mauritius, Rodriguez, and Reunion Islands of the Mascarene group in the Indian Ocean. These hyperspecialized and thus defenseless creatures were related to the pigeons. Pigeons are among the world's greatest fliers, but the island birds went their evolutionary way in the direction of sheer size, while their wings atrophied. They were no match for the European sailors who inevitably arrived on their islands, and theirs were the earliest extinctions for which dates are available.

Mauritius and Reunion were "discovered" by European sailors in the first decade of the sixteenth century, at about the time of Balboa's Panamanian success. The lamented dodo of Mauritius had become extinct by 1681. The Reunion solitaire lasted until 1746, and the solitaire of Rodriguez until 1791; but the first appearance of man on their islands signed and sealed their sentence. It has been the same world-wide. Since Columbus chanced upon the West Indies, some forty-six forms of birds and mammals have become extinct in those islands. At least forty-one more are on the critical list today.

All of them would still be thriving, had there been no navigation. The age of exploration ranks with the period of Pleistocene overkill and the current technological orgy as a land-
mark in the impact of man on wild nature.

When we recognize the immense self-confidence of
Renaissance man, his new navigational aids, gadgets such as
the fourteenth-century Chinese import, gunpowder, and his
growing efficiency in the applied sciences, it is not surprising
that when his self-esteem was reinforced by religious orthodoxy, he set about spreading that orthodoxy. If necessary,
he would spread it by force.

Whether mercantilism or religious conquest was foremost
in Europeans' motives, it is difficult to say. The upshot was,
however, especially in the Catholic seafaring countries, a zeal
for subdual which has only been equalled since in the invasion
of the moon. Might was inevitably right.

There is colossal arrogance in the phrase "Age of Discovery." In the sixteenth and the seventeenth centuries, Europeans "discovered" not only new lands, new faunas, new animals and plants, but whole civilizations. The assumption was
that these phenomena had been placed there, by almighty
providence, for the discovery, edification and profit of Christian Europeans, and of the institutions they represented.

At this distance in time, the events which followed the
1519 landing of Hernan Cortes at the beautiful and tranquil
island of Cozumel seem incredible. Yet it is possible to find
a modern counterpart. Rolf Edberg reminds us that "In Mon-
tezuma's capital city, Cortes kills as many people in the name
of His Highest Catholic Majesty as atomic bombs were later
to sweep away at Hiroshima and Nagasaki." But Cortes did
not have nuclear fission. No one knows how many Indian
people died in the course of the Spanish take-over, because
it is not certain how many there were beforehand. There have
been recent estimates of perhaps 40 million for Latin America
before the conquest. For Mexico, it is reckoned that the population at the time of Cortes' arrival was about 11 million. By
the end of the century, it had dropped to 2.5 million. Another

estimate has put the population of central Mexico at 25 million pre-Cortes, with a drop to a little over one million by the end of the century. Whatever the absolute figures, the depopulation was unprecedented, save perhaps for the effects of the height of the Black Death in Europe.

Butchery on this scale would seem to be beyond the physical capacity of even the most vigorous will. The invaders had something else working for them, what J. H. Elliott calls "that grimmest of all *conquistadores*" – smallpox.

An introduced disease, like any introduced organism, can play havoc with people – or other animals – which have not been conditioned to living with it. Just as introduced rats, goats, swine and other satellite animals of man disposed of the unique faunas of so many oceanic islands, smallpox (and measles) disposed of vulnerable people who had not lived with the diseases long enough to have developed a natural immunity to them. There is also evidence that following the ravages of smallpox in Mexico "social disorganization seems to have paralyzed recovery" (Ohlin). As in all natural situations, a homeostasis or equilibrium exists in human and other animal societies. It has been arrived at over many thousands of years of slow development. Sudden and widespread change, introduced over a short period of time, can destroy a society and reduce its chances of recovery to the point of no return.

The Spanish Christians happened to be Catholics, but the Catholics did not have a corner on vengeful and fiery retribution for the heathen. Protestants were equally savage, and, with the Catholics of the sixteenth century, believed that the proper penalty for heresy was burning. Protestant or Catholic, the Christian of the time viewed the non-Christian "pagan" as a kind of savage creature, a species of "animal" whose welfare was thus unimportant. While the heathen was outside the persuasion of Christianity, he did not exist as a human being. And if he was not a human being – he was nothing. It follows that the Christians' attitudes toward animals and wild nature were accordingly arrogant or indifferent.

W. H. Prescott revealed an interesting difference between the two Christian beliefs. He noted that the Catholics felt a duty to fight the battle of Christ in order to rescue pagan souls for eternity. The Protestants, perhaps even more egocentric in their view, considered that "Providence never designed that hordes of wandering savages should hold a territory far more than necessary for their own maintenance, to the exclusion of civilized man." Today, civilized men of all faiths hold true to that view, as any Eskimo can attest.

Trevelyan, in speaking of Drake, put it another way. Drake "wanted no inch of Spanish soil in the old world or the new. His objects were booty, trade, freedom to sail the seas and worship God aright, and ultimately to colonize empty lands where the Red Indian nomad would be the only person aggrieved." All of which, for both Spaniard and Englishman, reflects the colossal insensitivity of Christianity to the non-Christian world, whether it be a world of human or non-human beings.

The traditions of Western ethnic strains in relation to nature, wildlife, and conservation are intriguing. There is a pattern of distribution of the "nature ethic." Though the tradition is insufficiently developed anywhere, there are recognizable differences. The "best" of conservation awareness and practice is found in those parts of the world which have been influenced by Scandinavian, Anglo-Saxon, and German traditions – in general, the cultures of northern Europe. Also in general, the "worst" of conservation awareness and practice is conspicuous in those places largely dominated by French, Spanish, Portuguese and Italian traditions and cultures.

There are exceptions but, on a broad canvas, the breakdown is clear between northern Europe and the Romance traditions. There is a parallel contemporary difference between Protestant and Catholic traditions. It is no accident that the modern bullfight – surely one of the more horrendous manifestations of human attitudes toward nonhuman beings – continues in those countries where it does.

The appeal of the bull-ring was expressed by Ernest Hemingway, an *aficionado* for his own reasons: "A growing ecstasy of ordered, formal, passionate, increasing disregard for death. . . . It is impossible to believe the emotional and spiritual intensity and pure, classic beauty that can be produced by a man, an animal, and a piece of scarlet serge draped over a stick" (*Death in the Afternoon*). There are interesting insights in that passage – spirituality and, of course, the separation of man and animal. Christian spirituality is often as bound up with conquest as it is with the man/nature dichotomy.

It is difficult to detect any significant difference between the missionary zeal of the sixteenth- and seventeenth-century conquerors and that of their twentieth-century counterparts. There is little discernible difference between the invasion of Central and South America and the invasion of the moon, or of the arctic tundra. All represent the advancement of a faith – and it is in man's most radical being to stop at nothing in this pursuit.

CHAPTER TWELVE

The Great Ego Trip

Man is the introspecting animal. The possession of self-awareness and self-consciousness is included among definitive descriptions of our species. If there is one thing that distinguishes man from other animals, it is his mind. In it dwell thought and feeling, consciousness and volition; as St. Thomas Aquinas saw it, "those faculties of the soul which in their operation dispense entirely with matter." Wordsworth proclaimed the mind of man a thousand times more beautiful than Earth itself.

Of all his firmament of attributes, of all the arsenal of his skills and potentialities, none is so important to man as his Reason, which a generation after the death of Christ, Epictetus held "shapes and regulates all other things." Somewhat later, George Bernard Shaw was to observe, "The man who listens to Reason is lost; Reason enslaves all whose minds are not strong enough to master her." Reason is heady, and at least occasionally intoxicating.

The worst of Aristotle – his man-dominated ladder of increasing perfection – was rationalized with traditional Judaeo-Christian dogma by St. Thomas Aquinas. After Aquinas, the anthropocentric intellectual "construct" was elaborated upon significantly by Francis Bacon and René Descartes.

It was Bacon's creed that "knowledge is power." Knowledge is the "perpetrator of man's empire over the universe." Bacon is the patron saint of the technocrat. His unshakable belief was that "the true and lawful end of the sciences is that human life be enriched by new discoveries and powers." Always powers.

Bacon's preoccupation was with causes. He observed an effect, then worked back toward the cause or causes. He would then experiment to see whether his discovered cause produced the observed effect. It was an ultra-mechanical approach which he would have applied, had he lived, to *all* things – not merely the sciences. Thus explained in the second (1620) of his anticipated six books, Bacon's method was to be applied "to all the phenomena of the universe." It is still the human assumption that Baconian methodology, through sheer Reason, can be applied to all the phenomena of the universe. With Bacon, the Great Ego Trip was under way.

Bacon was very much a theist, which fitted nicely with his highly evolved homocentricity: "They that deny a God destroy man's nobility, for certainly man is of kin to the beasts by his body; and, if he be not kin to God by his spirit, he is a base and ignoble creature." The separation of God/man/nature was fundamental. In *The Advancement of Learning*, Bacon observes that "when the mind goes deeper, and sees the dependence of causes and works of Providence, it will easily perceive, according to the mythology of the poets, that the upper link of Nature's chain is fastened to Jupiter's throne."

The Great Chain of Being, as Arthur O. Lovejoy characterized it, is a notion that is pre-Aristotelian. Like all the other conceptual pictures of life and the universe, it proceeds in linear progression from beginning to end. The end is sometimes man, sometimes God, but man is always given superiority over the rest of nature. The chain analogy is of course the source of the irritating and quite erroneous notion of the "missing link," which implies that there can be, and was, a definable stage between *Homo sapiens* and his predecessors. Far from being an orderly determined progress, evolution is a three-dimensional searching and probing process with events occurring not sequentially but simultaneously. The direction of evolution is toward diversity, not linearity. It is quite obviously satisfying to the collective human ego, however, to see a long and orderly succession of linkages terminating with ourselves, or with a god in the anthropomorphic image. The Great Chain of Being, like Aristotle's ladder and Teilhard's tree of life, is a comforting and reassuring picture.

Garrett Hardin describes the Great Chain of Being: "All living organisms represent stages in the idea of 'Being,' each intermediate form being a more advanced stage than some, and a less advanced stage than the rest. If we had before us all the organisms that have ever existed, we could arrange them, single file, in a continuous chain of increasing complexity." For the atheist, the last link is man; for the theist, God. Either way, man's superior position is conspicuous by comparison with those of lesser creatures.

It was Francis Bacon's belief that inquiries into scientific and religious matters should be kept separate. For him, philosophy has three objects: God, nature, and man. They should not be confused, because they are discrete. Later in the seventeenth century, science and religion were combined in such ways as John Ray's 1691 *The Wisdom of God Manifested in the Works of the Creation.* For Bacon, however, God had provided two quite separate and different avenues of approach to revelation – the Bible, and the physical world.

Bacon made science "respectable," in the sense that during the Middle Ages science had been mostly restricted to alchemy at best, black magic at worst. Science had been widely considered profane. Bacon sanctified science, which now became the revelation of God's acts, and "the exaltation of God's glory."

A generation younger than Francis Bacon was René Descartes, the greatest of all apostles of Reason. So confident was Descartes in the reasoning capacity of the human mind that he felt it would be possible to discard all books and then to reason the universe into understandable terms. He rejected most scholastic tradition and all theological dogma. But it seems that he threw out the baby with the bath-water when he also disposed of all the joys of intuition. Bacon had been confident of man's ability to Reason and thus to "end in certainties," but Descartes was downright vainglorious. For him, knowledge would allow men to become "the masters and possessors of Nature."

If Bacon was the first of the modern scientists, Descartes was the first of the modern philosophers. His *Discours de la Méthode,* published in 1637, is a detailed account of his long and passionate affair with his own mind. He refused to believe anything in books unless it was supported by what he called "incontrovertible and absolute" proof. Descartes soon discovered that the only thing that could stand his test of doubt was the fact of his own existence. Thus, *"Cogito: ergo sum."*

For Descartes, it was vital to separate mind and matter. All things fall into one of these two categories, and neither can influence or benefit from the other. By separating mind and matter, Descartes was also able, as at one stroke, to separate religion and science, and to allow scientific inquiry to proceed without the limiting inhibitions of dogma. There was no doubt whatever in the mind of Descartes that all things can be discovered and known by the Rational Soul. As he says, " . . . all things, to the knowledge of which man is competent, are mutually connected in the same way, and there is nothing so far removed from us as to be beyond our reach, or so hidden that we cannot discover it. . . . " Descartes thus provided another rocket stage for the Great Ego Trip. We were ready for Descartes, and we embraced him. He was properly sanctimonious, he was properly scientific. He had further sanctified uniquely human Reason, and thus man.

At the same time, in another part of France, there was Blaise Pascal, who believed that "to think well is the principle of morality." Pascal – brilliant, religious, mystical, and "sublimely" articulate – was possessed of a humility which was not common in the Age of Reason. "The eternal silence of these infinite spaces (the heavens) terrifies me." He also acknowledged that "the whole visible world is only an imperceptible atom in the ample bosom of nature." Such remarks prompted many of his contemporaries and successors to brand him as a pessimist! As a man, he could not of course move from his position of uniqueness, but that limitation was not unique to him, nor to his time.

And then there was Thomas Hobbes – materialist, mechanist, political theorist. Like Bacon, he was persuaded of the immediate and future practical values of science and knowledge. Like all the others, he separated man and nature, and treated these in turn separately from supernatural inquiries. With Alexander Pope, he believed that "the proper study of mankind is man," and got on with his rationalist materialism somewhat to the dismay and resentment of the religious. For Hobbes, Reason is the covenant that binds together human societies, for without Reason man is "poor, nasty, and brutish," motivated entirely by appetite and desire, with his natural state being constant strife, enmity and war. Only Reason is his salvation.

Despite Hobbes and the "killer ape" theory and the "fall of man" notion, our primate roots are not in strife, enmity and war. Rather, they are in tranquillity and inoffensiveness. Hobbes' was a sterling defense of the advancement of science, but perforce it carried forward with it the tacit assumption of the uniqueness of man and Reason in the universe, and thus the "superiority" of man over nature. Nature was in no way permitted to intrude upon his thoughts except as the raw material for the advancement of human goals.

No one doubted that man's essential nature was his rational soul. Scientists and philosophers, whether religious or not, had Reason as their common bond. Newton, a strangely religious man (and a practicing alchemist) was, like Pascal, possessed of an aberrant humility: "I do not know what I may appear to the world, but to myself I seem to have been only a boy playing on the sea-shore, and diverting myself in now and then finding a smoother pebble or a prettier shell than ordinary, whilst the great ocean of truth lay all undiscovered before me."

One can only surmise on the extent of the anguish which may have been in the mind of a man who had such conflicting theological and scientific views. It is known, however, that it took the greatest tact on the part of his great friend Halley to persuade Newton to publish his *Principia*. Much of it would probably have been suppressed by Newton; he feared a religious controversy. A similar struggle with conscience was to be the fate of Charles Darwin, brooding in his study for so many years during the writing of *The Origin of Species*, never becoming entirely emancipated from the religious tradition of his childhood. Darwin felt as though he was "committing murder" by even entertaining the thought of a mechanism for evolution, much less publishing it.

During the golden period of evolving science, the thread which ran throughout was Baconian causation. Newton did not know how to handle the ultimate causes or origins of things. He had postulated a cosmic mechanism that operated according to universal laws. It did not require constant force or power, for it was self-propelling. God was needed only to start it. But just in case some constant power source *were* needed, Newton, in Bernal's words, "left a loophole for divine intervention to maintain the stability of the system."

Newton, a Platonist, believed in a Supreme Being, and thus had to allow that Being a role in his cosmology. (It was in this way, it is said, that Platonism entered the mainstream of our Western thought – by way of Newtonian physics.) Newton, who was sufficiently honest to admit his ignorance about original causes, nonetheless believed in the ultimate *necessity* of original causes, and had to leave things to the will of God at the beginning of Creation. Mechanistic scientists have always been embarrassed by this, but it is difficult to fully understand their embarrassment. If there *were* beginnings of the cosmos, there is no evidence to refute that they were the will of God. In the framework of the cosmic instant which will be the career of the species man, original causes seem scarcely relevant.

Reason presupposes causes, and causes presuppose necessities. The occasional dissenter, such as David Hume, has been largely ignored. Hume denies causes and effects: "Necessity is something that exists in the mind, not in objects." He also denies any necessity for "self." He feels that the pursuit of Reason has succeeded only in bringing us "into a frame of mind where the solid fabric of the world dissolves like a dream before our eyes, or passes into a kaleidoscopic unreality of change."

Human Reason has had a long history, and the Great Ego

Trip is far from over. We have made it to the moon, and we have sent a space vehicle to circle Mars, and photograph it. Such notable accomplishments add immense impetus to the self-glorification of the reflecting primate.

Primates as a group are "generalized" creatures. They can eat almost anything within broad limits; they can live virtually anywhere. Thus, they have been so successful over tens of millions of years. Man has parted company from the other primates in only one major way – the development of his brain. In this sense, man is no longer generalized but highly specialized.

There are other animal species which have large brains, notably whales. We do not understand the brain of a whale, but experiments with dolphins and other smaller whales reveal that there is a good deal more cerebration going on than we anticipated. Dolphins and others can be "taught" – or encouraged – to do remarkable things, and when one watches them for a time there is no doubt whatever that they are capable not only of reasoning but also of "feeling."

It is strange, given the size and complexity of their brains, that the largest whales are so pathetically easy to kill. The much smaller and less convoluted brain of a crocodile or alligator allows the animal to quickly learn about people, jacklights and guns. Ducks and geese know where shooters lurk. Why is it then possible for a whaling fleet to wipe out whole pods of whales, one after another? These are thought to be the most intelligent nonhuman animals on Earth. Why do they not profit by observing the fate of their kind?

The largest whales are at the top of the ocean food-chain; they are the great grazers of the oceans. They have few, if any, important natural enemies. Occasionally, killer whales and some of the biggest sharks may take their newborn young, but this would not seem to be significant in terms of total numbers. Perhaps whale brains, because of their possessors' almost absolute security, do not need to conceive of fear. Perhaps, in its innocence, a whale can only conceive of love. There are many fully documented observations, both in the wild and in captivity, of whales such as dolphins which have demonstrated, not only to their own species but also to *other* species, that trait we call kindness. They have helped other kinds of whales, and they have also helped people. Perhaps the brain of the very largest whale is *over*specialized toward kindness and altruism, in that it cannot comprehend the notions of fear, or mistrust – or evil. Overspecialization, either behavioral or physical, can carry the seeds of eventual extinction. An animal with a peculiar and overdeveloped speciality, such as complete dependence on eucalyptus leaves, is surely doomed if eucalyptus leaves for some reason are not available. An animal with any highly developed speciality is doomed if the environment should suddenly change – perhaps ever so slightly – making that speciality no longer appropriate, or even advantageous. It is quite possible that the human brain may be such an overspecialization. Reason may be such a liability.

A strong case may be made for Reason as an overspecialization. Reason gave us the technology which is killing other species of animals and plants, killing us, and killing planet Earth. Reason allowed us to rationalize the divine mission to subdue. Unlike the blue whale, Reason gave us the capacity to conceive of evil. Reason gave us the concept of species hierarchies and species dominance, a perversion of the natural social dominance of other primates.

In the sixteenth century, Montaigne reflected that science without conscience is but death of the soul. Montaigne was no technocrat: "Let us a little permit Nature to take her own way; she better understands her own affairs than we." Montaigne, like Hume and Pascal, represents the positive side of Reason – the warmth of wisdom.

As actions have their reactions, so emotions have their counter-balancing opposites: love/hate, hope/despair, joy/anger. Neither side of either pair can exist without the other. Such are the specializations of the human brain which have evolved side by side, simultaneously and concurrently, through time. They have been there as long as our brains have been human. The moment we could love we could also hate, and the moment we knew despair we were expressing the capacity for hope.

The evolution of Reason must perforce have involved

both sides of the old combinations – positive and negative. We could not have had Renaissance cathedrals without the slaughter of Aztec and Inca innocents; both were in celebration of the European church. We could not have had the advancement of science without the intellectual arrogance of the Age of Reason. These are the "trade-offs" which occurred along the ancient path of evolving culture. They are comparable to the inevitable trade-off the seal made, for example, when for success in the water it was compelled to forsake the ability to walk.

Reason is rather like a "technological fix." If it gets us into trouble, we attempt to apply just a little more of the same in order to extricate ourselves. Thus, frequently, we compound our problems. It is quite possible that this is the evolutionary future of our personal overspecialization. Like a peacock's tail, it will continue to grow, whether we like it or not (and, like the peacock, we are terribly proud of it), and it will reach a point at which it will no longer have survival value. From then onward, extinction is swift. Perhaps, with mindless and directionless technology, we have already passed the point of return – the achievement of humility.

Garrett Hardin takes a more optimistic view, arguing that intelligence is adaptive and thus unspecialized. "It is not like an instinct, finely adapted to an immediate solution to one problem. It is more a sort of skeleton key, capable of opening a wide variety of doors. But with it man may have opened one door too many and thus have led the way to his own extinction. We don't know. But if he does not kill himself, it will be because his godlike intelligence enables him to find but one more door."

The extinction of man is as certain as his existence. Extinction is certain for all species, faunas, planets and stars. Individual species, like individual beings, do not *matter* in the larger system of things. No individual or species will be selected for immortality. We will be mourned by no one. In Darwin's words – the world will go cycling on.

CHAPTER THIRTEEN

The Flutes of Arcady

One cannot be certain whether man is the only animal capable of nostalgia, but having in mind his well-developed capacity for self-deception, one might suspect that he is. This failing has been illustrated in religions and schools of thought from Aristotle to Teilhard. The latter, with only the best and loftiest of motives, was incapable of avoiding the ultimate self-deception – the position of man on the "tree of life."

R. D. Laing says, "Human beings seem to have an almost unlimited capacity to deceive themselves, and to deceive themselves into taking their own lies for truth." We do this not only as individuals, but also as groups, as societies, as cultures, and as a species. Man is the self-duping animal; in this respect at least, man is singular. This is especially true of nostalgia; things past which linger so glowingly in our remembrance rarely if ever existed – at least not in the forms in which we remember them.

Romanticism is nostalgia for things which never were. It is the ultimate in self-deception. We are all guilty of it, and at one time or another we all *know* we are guilty of it, but we persist in our romanticism. And, of course, we cannot face the possibility that there may never have been anything there in the first place. (That would be like denying that there is any Reason for things.) As Laing says, "Take anything, and imagine its absence." That is extremely difficult to do – if not impossible – because it is frightening. It is hard on our security. Laing adds the devastating thought, " 'There's nothing to be afraid of.' The ultimate reassurance, and the ultimate terror." *Nothing?* Just as we are "afraid to approach the fathomless and bottomless groundlessness of everything," we cannot face the fact that there never was an Arcady.

One knows there was a place called Arcadia. It was not, however, Camelot. Arcadia was a plateau on the Peloponnesus which was cut off by chains of mountains from the rest of what is now Greece. Thus, its people did not mix much with those of the other countries around it, and did not acquire elements of their cultures. It is said that the people and culture of Arcadia "stagnated." They remained hunters and pastoralists, while their more advanced neighbors in adjacent states, such as Sparta, were busy making wars.

At some point, perhaps even before Vergil, Arcadia became idealized as the never-never-land of total human happiness – even beatitude. Idyllic conditions were imputed to it, and the carefree shepherds and shepherdesses became the envy, the delight, and the concreted nostalgic remembrance of a large part of the Western world. The likes of Cowper, Tennyson – even Goethe – recalled Arcadia and its joyous, childlike innocence.

There have, however, always been heretics among us – people who knew that *You Can't Go Home Again;* but most of us have preferred to bask in the remembered joys of sunny, bucolic Arcady. Like fantasying of any kind, romanticizing the past is inconsequential and relatively harmless. It becomes important and potentially destructive when, through romanticism, we impute to the nonhuman world uniquely human emotions, characteristics, and attitudes. Disney romanticizes nature. It is a very fine line between romanticism and anthropomorphism.

The poet who sees benignity and benevolence in nature – or harshness and cruelty – is denying the utter indifference of nature to human existence and human values. Natural objects and nonhuman animals know neither nastiness nor saintliness. (I have long suspected that the Romantic poet, rejoicing in the lily, is somewhat narcissistically rejoicing in his sensitivity to the "lower" forms of life.) The Romantic does not appreciate Alexander Pope's "Whatever is, is right." All must be seen in human terms, and all must be possessed of human values. Otherwise, it is not real.

Man is obsessively preoccupied with rank. Man is the ranking and grading animal. Modest men may posit angels and God; immodest men may see the final link in the Great Chain of Being or "inflorescence" as man. Either way, *verticality* is inherent and essential; man is man, and after man there are all the "lower" species. All things must be classified, and all must be labeled. All must be given their appropriate position in file, whether by God or by man, in an immutable hierarchy. We have never learned to think in more than two dimensions.

Perhaps our preoccupation with hierarchical structures is a demonstration of Alfred Adler's concept of our constant and life-long battle with inferiority. Robert Ardrey relates this to the struggle for dominance, the juggling and jockeying for position and status. Perhaps Adler's phenomenon also encourages us to picture ladders, chains, and ascending orders of perfection, from Omega to Alpha.

We know that other primate societies, such as those of baboons and chimpanzees, have dominance hierarchies. Such structures are necessary for societal survival. There is no difficulty in accepting that hierarchical thinking is an integral part of our primate heritage, and that nothing in our makeup is more natural than to attempt to dominate other members of our society. But man extends this drive not only to individuals of his own species, but to *other species,* a phenomenon unheard of in nature. If we are to acknowledge, as we must, that there are Alpha individuals, then it seems logical to our illogical minds that there must be Alpha *species* also. Human thought is riddled with this type of syllogistic reasoning. Adler's picture (and a perfectly accurate one) of the *intra*specific individual power struggle becomes Reasonable man's picture of *inter*specific dominance hierarchies, which is a false and unnatural one.

In eighteenth-century Europe the whole concept of "Nature" underwent surprising evolutionary development. Setting aside for the moment the self-centeredness of the human position on the Great Chain of Being, the notion of a chain had promoted a new idea of unity in natural systems. All things were tied together in a "master plan," as though it were one great piece of machinery, operated by divine benevolence.

Thus we had the "divinization of Nature," as expressed in

the view of the third Earl of Shaftesbury, who, like Newton, was a deistic Platonist out of Cambridge. " 'True religion' should be based on 'Nature' rather than on Revelation . . . also, orthodoxy is the enemy of 'true' ('natural') religion, because it invites us to base our faith, not on the beautiful and harmonious Order of Things – the best and only genuine external evidence – but on miracles, that is, on infractions of that Order" (Willey). This, of course, was a bit much for the church to swallow, what with its traditional insistence upon revelation, but it was just going to have to manage. "Nature" was on the upswing.

"Nature" made it possible for a man to be religious whether he believed in God or not. The godly man saw nature as God's handiwork, and believed that God, "instead of expressing his will directly, was bringing it to pass through the realm of nature and natural law" (Soule). Those who had dropped religion were able to substitute the entire world of living nature. Those who had become disappointed or frustrated with man and his works were also able to substitute nature as the recipient of their loyalties. There was never any question, however, about who was the sole beneficiary of the grand design – Enlightened Man. Neither atheist nor churchman had any doubt about that.

There appeared, however, a growing body of people who were disappointed and frustrated by man, and angered at the accelerating expansion in human self-esteem. The widespread acceptance of the new religion of Nature, especially as Nature was revealed and understood through Reason, made this period an ideal climate for the emergence of a group of satirists, who assailed eighteenth-century civilized European man for his overweening "pride." The smile of Reason has spawned the vice of pride, and Pope was vigorous in his reactions:

In pride, in reas'ning pride, our error lies;
All quit their sphere and rush into the skies!
Pride still is aiming at the bless'd abodes,
Men would be angels, angels would be gods.

This was a reference to a certain amount of jostling for position which Pope had noted at the top end of the Great Chain of Being. Pride, the product of misused Reason, is throwing a spanner into the perfect machine.

> All nature is but art unknown to thee.
> All chance, direction which thou canst not see;
> All discord, harmony not understood;
> All partial evil, universal good;
> And, spite of pride, in erring reason's spite,
> One truth is clear, Whatever is, is right.

No criticism of the time was so savage as that delivered by Dublin's Jonathan Swift, whose moral indignation and sincerity of purpose drove him to attack not only politicians and prevailing social customs, but also the entire human species. For Swift, man has betrayed his own Reason, and through sheer stupidity and vanity has distorted his own values and even his own judgment. Swift reverses the traditional hierarchy and places the filthy, degenerate Yahoo at the bottom of the ladder. As Lovejoy describes it, "The most detestable and irrational of beings (man) crowns his fatuity by imagining himself the aim and climax of the whole creation." Two centuries later, man is still regarded as the aim and climax of the whole creation.

Anyone who has spent the greater part of a lifetime enjoying and attempting to understand and preserve wild nature will have had the experience of witnessing his own species drift lower and lower on his personal scale of perfection. All the magnificence and nobility of our creativity cannot begin to compensate me for what my species has cost me. Shakespeare cannot compensate me for toxic pesticides, Bach cannot compensate me for Hiroshima, nor Michelangelo for the blue whale. Jesus Christ cannot compensate me for the brutal imposition of human power over nonhuman nature. Yet, the total destruction of blue Earth may well precede any diminishment in human pride.

Misanthropy is probably the inevitable (or at least occasional) companion of the man who values unspoiled nature, natural systems, and their individual components. Byron could say, "I love not man the less, but Nature more," and could steal moments to "mingle with the Universe." He was aware that "Man marks the earth with ruin."

There was also a great wave of nostalgia. But it was not nostalgia for any early time that had actually existed; it was nostalgia for the word-picture of Arcadia. It was Romantic. In John Ruskin's words, the nostalgia was a "pathetic fallacy," for the new movement saw in Arcadia, and in nature, things that were never there. The Romantic imputes to nature emotions that are peculiar to the Romantic, not to nature. He is recasting nature in his image; he is anthropomorphizing it.

Feelings such as pleasure and grief are not unique to man. Nonhuman animals show affection, and they show sadness; but they do *not* attribute the capacity for these emotions to other animals, nor do they read mirror-pictures of themselves in the behavior of other animals. That is a solely human characteristic, so far as we know. Animals do not zoömorphize, probably because they cannot conceive of interspecific dominance and thus have no need to recast the rest of the world in their own image.

Nostalgia generally involves a sweet kind of sadness, and there was plenty of melancholy in the Romantic movement. Jean-Jacques Rousseau and his followers were thoroughly disenchanted with the state to which men had brought eighteenth-century European society, and they bewailed the fall from grace that was represented by man's departure from a "state of nature."

Rousseau had great difficulty in defining a "state of nature." Certainly it could not be the "brutish" situation described by Hobbes. That would have been unthinkable to the European mind of the time. But civilized man, in Rousseau's view, was even worse than Hobbes' beastly "primitive," which was little more than a Yahoo. So he was forced to make a compromise. As Leo Marx says, "Unable finally to endorse either the savage or the civilized model, Rousseau was compelled to endorse the view that mankind must depart from the state of nature – but not too far." Marx goes on, "Curiously enough, Rousseau thought that mankind had passed through the ideal state during the pastoral phase of cultural evolution, by which he meant a pastoral situation in a literal, anthropological sense: a sense of herdsmen." Rousseau believed not in original sin but original good. Man the pastoralist was seen as man in nature, and that is where he should have stayed.

Rousseau's arrival at this position was, ironically, a prime example of the use of Reason in rationalizing a highly complicated point of view. Ardrey has been hard on Rousseau for this. "The organizing principle of Rousseau's life was his unshakable belief in the original goodness of man, including his own. That it led him into the most towering hypocrisies . . . is of no shaking importance; such hypocrisies must follow from such an assumption. More significant are the disillusionment, the pessimism, and the paranoia that such a belief in human nature must induce."

Surely, however, the assumption of original good is no worse than the assumption of original evil, or of any other starting point. It is now in our tradition, even as we move toward the final quarter of the twentieth century, that there have always been beginnings, and causes, and directions. It is difficult to single out the eighteenth century for castigation on the basis of such assumptions. Hobbes was just as wrong as Rousseau, and the twentieth century is just as wrong as the eighteenth.

Romanticism is a great deal easier to accept than Reason. Romanticism at least sees some value, even if it is only a human-centered value, in the nonhuman world, and admires things which are not the products of Reason. The longing for Arcadia is pathetic, but the worship of human intellect to the exclusion of sensibility is preposterous and potentially suicidal. Daniel McKinley has suggested the "possibility of a core of sense in what may have seemed just another fad," because romanticism may have "generated some salvageable truths that previous centuries had failed to disclose. . . . Romanticism, in its reverence for the integrity of the exotic and its presuppositions of a divine fullness in kinds of things, helped lay the groundwork for modern ecology and natural history."

There is, however, a danger inherent in Romanticism. The nostalgic view of nature, with man set in it in the form of a gentle pastoralist, does just as much to separate man from his environment as does the cult of the cold intellect. Both are anthropocentric. Both see nature and man as separate entities, albeit in different relationships. The Romantic of Rousseau's school sees man using nature as a plaything or as a backdrop for the display of man's most admirable characteristics. Nature is the mirror for this self-admiration.

Irving Babbitt has made this point with conviction. "The Rousseauist does not in his 'communion' with nature adjust himself to anything. He is simply communing with his own mood. Rousseau chose appropriately a title for the comedy that was his first literary effort *Narcissus or the Lover of Himself.* The nature over which the Rousseauist is bent in such rapt contemplation plays the part of the pool in the legend of Narcissus. It renders back to him his own image. He sees in nature what he himself has put there. . . . Nature is dead, as Rousseau says, unless animated by the fires of love."

The Romantics were forever moralizing about the natural world; they were determined to find "good" in nature, and to show how contemporary man was "evil" by comparison. By extension, Romanticism in the "Nature" context becomes akin to religion. It comes very close to being institutionalized, complete with dogma. This seems to be the inexorable evolutionary trend of almost any body of thought, or even of feeling. All must be codified. Nature is virtuous; ergo, man is unvirtuous, because man has left nature. The separation between man and nature is categorically described, as it is in our organized religions. Like any other creed, Romanticism offers simultaneous threats and promises, giving with the one hand and taking away with the other.

Granting the best motivations to Rousseau's cause (with which any naturalist can only sympathize), it points up a fundamental characteristic of mankind – the compulsion to forever *evaluate* things, to *pass judgment.* This compulsion seems inescapable. Men of Reason pass judgments, as do Romantics. Scholars moralize, as do poets, escapists, misanthropists, conservationists and technocrats. We are driven by the urge to categorize things. And more often than not, we use the age-old scales of perfection. This is better than that, and that is worse than this. And we recommend cures for erring human nature. Finally, we evaluate all things only in human terms, and by human standards.

Even our language is constrained by the subjective judgments inherent in the words we use. Take that innocuous word "humane." The very best of dictionaries defines "humane" as civil, courteous, obliging, kind, benevolent; it evokes mercy and compassion. These are the properties inherent in the root word "human." On the other hand, we find no mention of the rest of the long list of uniquely human attributes so painstakingly spelled out by Hobbes, Swift, and even Ambrose Bierce. Where are uncivil, discourteous, selfish, cruel, malevolent; and where are merciless and pitiless?

We have amputated one-half of the human equation and expunged it from our language. It is no wonder we have such great difficulty in defining "human nature"; we no longer have the words for it, unless we look up "paradoxical" or "schizoid," or some such. The Romantics saw only one side of human nature, the misanthropists only one, the men of Reason only one. Has any man been able to see all sides at once, save the rare likes of a Shakespeare or a Pascal? Perhaps only the cubist painter is capable of that.

It is a melancholy fact that many of those who extol the virtues of nature are themselves seriously out of joint with their human society and surroundings. The aphorism "Nature is nobler than man" is just as fatuous as "What did nature ever do for me?" We have heard both points of view. I have heard the former from the occasional recluse who, because of a personality short-circuit, is estranged from the human community. He is afraid of people and he knows that a sparrow or a toad cannot hurt him. Like the pet-fancier (if Adler is right), he indulges his frustrated dominance drive in that non-human direction. Such a posture is pitiable.

The position "What did nature ever do for me?" is more than pitiable; it is destructive. Estrangement from nature is estrangement from oneself, because one *is* nature. One is also posterity. The publisher (or the chairman of the board) of a major U.S. popular magazine recently asked publicly, "What did posterity ever do for me?" The rhetorical question was raised in the context of some widely publicized environmental issue of the time. Such a statement is barely Neanderthal, but the business press is replete with them.

The Romantic movement was filled with affectation, posturing, self-conscious sensitivity and sensuality, misanthropy and empty nostalgia. Yet it revealed other ways of being human, of examining the phenomenon of man. The flutes of Arcady were never there, but then, neither was original sin. There was no "noble savage," there was no Yahoo. There was only introspecting, self-deceiving man, frantically searching for homocentric meanings, purposes, and designs, when all the time there were none. The enormity of that simple truth has yet to dawn on us.

CHAPTER FOURTEEN

Mission Accomplished

From the earliest days of North American colonialism, the mission appeared multi-faceted, but its essential purpose was profit. Only the anticipation of profit could have prompted the backers of the *Mayflower* voyage in 1620 to underwrite the transportation of 102 Protestant dissidents to the new land. (It is ironical that the Puritans, so vexed by religious intolerance in England, were to prove less than tolerant once in the New World.)

The spirit of John Calvin met and conquered the immense North American wilderness out of sheer perseverance – and the sense of mission. Calvin was an indomitable man, the model of tenacity. He was also possessed of rare imagination. But even Calvin, at his death in 1564, could not have foreseen the colossal level of achievement which his model of dogged industry would inspire – the subjugation of an entire continent which had been revealed to Europeans only seventeen years before his birth.

There is little doubt that the Pilgrim Fathers and their financial backers had carefully read tracts such as Alexander Whitaker's *Good Newes from Virginia*, published in London in 1613. Who – most especially a mercantilist – could resist such intoxicating reports as: "Wherefore, since God hath filled the elements of earth, aire, and waters with his creatures, good for our food and nourishment, let not the feare of staruing hereafter, or of any great want, dishearten your valiant minds from comming to a place of so great plentie: if the Countrye were ours, and meanes for the taking of them (which shortly I hope shall bee brought to passe), then all these should be ours: we haue them now, but we are fain to fight for them, then should we haue them without that trouble."

The instigators of settlement in the new continent used every device, including artists and the church, to further their cause. As Louis B. Wright has put it, "A share of stock bought in a colonial company would be a stroke for both God and King, for the advocates of settlement in North America solemnly asserted their high purpose of saving heathen souls and establishing outposts against the extension of the Spanish empire." In 1622 there occurred an Indian massacre which laid waste a number of frontier farms, and the Virginia Company "persuaded no less a person than Dr. John Donne, the poet and dean of St. Paul's, to preach and publish a sermon declaring, somewhat metaphysically, that support of the colony would advance both England and the Kingdom of God. For this useful message, Donne received stock in the Virginia Company and for a time nourished the hope of becoming secretary of the corporation." Donne also contributed a short poem to accompany the publication of John Smith's account of New England, dedicated "To his friend Captaine Ion Smith, and his Worke."

The settlers were not long in noticing the stupendous abundance of American wildlife. There have been many published descriptions of the flights of passenger pigeons, but one of the earliest is in Wood's 1634 *New-England's Prospect:* "The Pigeon of that Countrey is something different from our Dove-house Pigeons in *England,* being more like Turtles, of the same colour; but they haue long tayles like a Magpie. And they seeme not so bigge, because they carry not so many feathers on their backes as our *English* Doves, yet are they as bigge in body. These Birds come into the Countrey, to goe to the North parts in the beginning of our Spring, at which time (if I may be counted worthy, to be beeleeved in a thing that is not so strange as true) I have seene them fly as if the Ayerie regiment has beene Pigeons; seeing neyther beginning nor ending, length, or breadth of these Millions of Millions. The shouting of people, the ratling of Gunnes, and peltime of small shotte could not drive them out of their course, but so they continued for foure or five houres together. . . . "

The Reverend Francis Higginson had already mentioned the great "Flockes of Pidgeons" in his 1630 *New-Englands Plantation. Or, a Short and True Description of the Commodities and Discommodities of that Countrey,* but he clearly found the wild turkeys more to his taste. "Here are likewise aboundance of

Turkies often killed in the Woods, farre greater than our English Turkies, and exceeding fat, sweet and fleshy, for here they haue aboundance of feeding all the yeere long, as Strawberries, in Summer all places are full of them, and all manner of Berries and Fruits." The English turkeys to which he referred were descendents of the same American species, brought to Europe from Mexico by returning *conquistadores* and introduced to Britain about 1540. Higginson also rhapsodized about the Massachusetts air, which must have been a godsend after the sulphur dioxide coal fumes of grimy London: " . . . a sup of *New-Englands* Aire is better than a whole draft of old *Englands* ale . . . "

Whether turkeys or pigeons, deer or bear, wildlife species were there for the consumption of the God-fearing settlers. John Josselyn, in *New-Englands Rarities,* which was published in London in 1672, provided a long list of American birds and mammals, and the uses for which they had been provided. One of the most interesting of these concerns *The Wobble.* "The *Wobble,* an ill-shaped Fowl, having no long Feathers in their Pinions, which is the reason they cannot fly, not much unlike the *Penguin* ; they are in the Spring very fat, or rather oyly, but pull'd and garbidg'd, and laid to the Fire to roast, they yield not one drop." The Wobble is the now-extinct great auk, which flourished in large numbers on its remote North Atlantic islands until Europeans arrived in sailing vessels. The flightless birds persisted until the 1840's, when the last individuals disappeared forever beneath the clubs of fishermen.

Josselyn had a medicinal application for great auks. "Our way (for they are very soveraign for *Aches*) is to make Mummy of them, that is, to salt them well, and dry them in an earthen pot well glazed in an Oven; or else (which is the better way) to burn them under ground for a day or two, then quarter them and stew them in a Tin Stew-pan with a very little water." He even had a use for the sharp, curved bills of ospreys: "Their Beaks excell for the Tooth-ach, picking the Gums therewith till they bleed."

Not all, however, was strawberries and roasted fowl. The land was hard. In the seventeenth century, the bleak Calvinist clergyman Michael Wigglesworth, author of *Day of Doom,* was moved to identify "God's Controversy with New England." Convinced that New England had been singled out for divine displeasure, the glum Wigglesworth reached back to Deuteronomy for his description of the land as a "waste and howling wilderness." This was to be the consensus for some time to come. But it was understandable. Peter Matthiessen has the pulse of the matter: "One must keep in mind that the glad reports of this period were motivated in part by the colonists' desire to share with others an existence made miserable as much by their own religious strictures as by hardship!"

Tight-lipped and unsmiling, the Puritans went about their business of conquest. The process was less spectacular than the blood-lusty deeds of the *conquistadores,* but the Puritans had come to stay, not to pillage. They had come to master, and it took a special kind of people to master wild North America. They had become that special kind of people long before they left England, because, as George Grant says, "When one contemplates the conquest of nature by technology one must remember that that conquest had to include our own bodies. Calvinism provided the determined and organized men and women who could rule the mastered world. The punishment they inflicted on nonhuman nature they had first inflicted on themselves."

Frontier people always fear and mistrust the wilderness, and usually they have good reason to. A subsistence existence is a precarious one, and even a relatively minor accident – a sprained ankle or a touch of frostbite – can have disastrous consequences. Considering the level of technology at the time, one can understand the pioneers' dread of the wilderness. However, the emotional evolution from fear to hate is less comprehensible.

Roderick Nash has identified what he calls a "tradition of repugnance" for the American wilderness. This tradition persists on the frontiers of our time. It is not easy to determine its origins. Certainly the Judaeo-Christian tradition is paternalistic and condescending toward wild nature, as well as aggressive, whether one chooses either the "dominion" or the "stewardship" interpretation of the Bible. But it is not openly hostile. Hatred of the wilderness as expressed by the colonizing Protestants seems to have been their own special aberration, without valid biblical precedent. Of course it is easy to read almost anything into many passages of both Testaments, and it would appear that the settlers seized on such texts as they needed to justify the holy war on nature.

The pioneers came to regard wild nature as not only dangerous but malevolent. Fortified by an obligation to dominate in God's name, the mission to destroy became obsessive. Thomas Merton has remarked that the settlers came to hate the wilderness "as a *person,* an extension of the Evil One, the Enemy opposed to the spread of the Kingdom of God." The Christian duty was to dispatch the enemy with speed and efficiency. The subjugation of nature became a crusade, prosecuted with ferocity.

The crusade required all that was best in the Calvinist character – grit, stamina, stubbornness, belief – plus lack of imagination and sensitivity. It required mission-oriented indifference to consequences, and zeal fired and stoked by God's will. Once it creaked into motion, the accelerating juggernaut was unstoppable. All gave way before it – pine and oak forests, grasslands, wildlife.

As Charles Darwin knew, "Ignorance more frequently begets confidence than does knowledge." Though the pioneers were prepared to endure severe hardship, they never questioned that they would prevail. The momentum of Right would carry them through, if they practiced diligence, thrift, etc.

There was a strange irony in this, described by Nash as the fact that "success necessarily involved the destruction of the primitive setting that made the pioneer possible." With the wilderness went the pioneer, and with the pioneer went the archetypic Puritan. A new North American emerged, characterized by an odd ambivalence about the land. On the one hand, the new master of nature carried the lingering heritage of hostility and contempt for wilderness, yet on the other he had imported from Europe a romantic vision of pastoral nature no less strange than that of Rousseau himself. This was a man seriously divided, and he persists to this day.

A splendid and penetrating analysis of the American schizophrenia about the land has been contributed by Leo Marx in *The Machine in the Garden*. Like European Romanticism, Marx says, the American pastoral tradition has man very much outside nature, but observing nature and deriving pleasure from it. "What is attractive in pastoralism is the felicity represented by an image of a natural landscape, a terrain either undespoiled or, if cultivated, rural." This is at odds with the dreadful desolation described by many of the Puritans, but Marx provocatively suggests that the image of hideous wilderness was exactly what the Puritans *wanted*. "Colonies established in the desert (as opposed to the gentle garden) require aggressive, intellectual, controlled, and well-disciplined people." The ugly picture of the wilderness was a challenge to the exercise of will and power, and it was a challenge that was eagerly met.

Throughout North American history, there has never been any reason for the existence of wildlife species save to serve man – whether for food, clothing, amusement, or profit. If a species did not serve human purposes, of what possible "good" could it be? If, as sometimes appeared to be the case, a species actually "competed" with human interests (too frequently, recreation), then clearly it must be eliminated. For some inexplicable reason, God, while creating doves, squirrels, and grouse, had also created coyotes, hawks, and rattlesnakes. This is what Jack Olsen calls "the Mother Goose table of animal values (wolves and bears are evil, bunny rabbits and chipmunks are good, etc.)."

Jack Miner, friend and patron of the Canada geese, called the great horned owl a "heartless cannibal," and made it his business to kill twelve to fifteen of the birds every winter. Divine wisdom had somehow slipped up in visiting such demoniac creatures on man's world of nature. But no matter. It was man's holy mission to clean things up.

The bison was obviously no cannibal, but he was there for human purposes; and anyway there were so many that they actually got in the way. Buffalo Bill Cody, who boasted of killing 4,820 bison in one twelve-month period, was the idol of rich and poor. He turned his personal legend into a traveling sideshow which earned him fame on both sides of the Atlantic. The Boone and Crockett tradition evolved into something close to sacrosanct, and the firearm cult is still very much alive today.

Killing for amusement is a deep-rooted human tradition. Early man was a hunter, and he became a good one. But when people began to settle in more permanent communities and undertook pastoralism and agriculture, the need for wild game was no longer significant. Except for those relatively few peoples who continued to hunt as a way of life, wild game ceased to be an important element in the human diet. In the New World, for a time game was a marketable commodity, and professional gunners profitably plied their trade until they ran out of wildlife to kill in the latter part of the nineteenth century. In our own time, however, it is difficult to understand why the pursuit of such wild game as remains continues to be carried on by North Americans of European descent, when the necessity no longer exists.

On a lovely African evening, Roger Peterson and I were driving our Land-Rover up the slopes of Mount Kenya on the way to our campsite. We chanced to pass the tents of a large, highly organized gunning safari. Spread round the campfire were the inert trophies – a pair of elephant tusks, a rhino head, the stretched hides of antelopes. That sight, with all its implications, reeks in my memory. In the twentieth century, large African mammals such as elephant, sable antelope, bongo, greater kudu, and many more, have been reduced by nine-tenths. The romance of Darkest Africa is deeply involved with killing, as is the romance of those North American frontiers that yet remain.

The Judaeo-Christian progress tradition has come to be intimately involved with the profit motive, which in turn is intimately involved with man's impact on nature. The economics we know are Western phenomena; it was Max Weber, the turn-of-the-century sociologist, who observed that Western civilization is unique in that it supposes that its cultural trappings have *"universal* significance and value." Today, Western cultural traditions are dictating not only the course and destinies of those nations in which they originated, but also those of the "underdeveloped" or "developing" nations, which are in the process of being encouraged – one might even say coerced – to aspire to those values. None of these is more important than the economic system which is based on perpetual growth in the production and consumption of goods. Most of today's "under-industrialized" nations have been persuaded that in order to achieve the highest level of success, they must adopt the traditional Western approach to trade and commerce, and all the radical assumptions which underlie that approach. This has global environmental repercussions, as a potentially suicidal drain on non-renewable resources.

Resources were considered to be infinite and inexhaustible when Adam Smith pronounced in his 1776 *The Wealth of Nations* that, assuming that the object is to increase the national wealth, "this object will be most effectually secured by perfect industrial liberty." The consequences of perfect industrial liberty of access to resources have become conspicuous in our time. Smith's pronouncement became dogmatized, and a new body of faith – a creed – had emerged. Leave industry to its own devices, and all will be well. Such truths are universal, said Smith.

The universal truth of *laissez-faire* found an eager and willing marriage partner in Calvinism, and although a long series of people had conspired to bring the two together, the most generally recognized final matchmaker was Benjamin Franklin. Franklin had inherited all the Puritan virtues, and although the strictly religious context had largely evaporated by his time, all the fundamentals remained. As Weber says, "Honesty is useful, because it assures credit; so are punctuality, industry, frugality – and that is the reason they are virtues." These truths, among others, were held to be self-evident.

Weber's thesis, subsequently elaborated in R. H. Tawney's *Religion and the Rise of Capitalism,* was that there is an intimate and inseparable relationship between the asceticism of traditional Calvinism and the emergence of modern capitalist institutions. He points out that when asceticism "began to dominate worldly morality, it did its part in building the tremendous cosmos of the modern economic order. This order is now bound to the technical and economic conditions of machine production which today determine the lives of all the individuals who are born into this mechanism, not only those directly concerned with economic acquisition, with irresistible force." Machine production, of course, is not unique to capitalism.

George Grant remarks that "the public virtues (Franklin) advocates are unthinkable outside a Protestant ethos. . . . In 1968 Billy Graham at the Republican convention could in full confidence use Franklin in his thanksgiving for what the Christian God had done for North America." The tradition of Dr. John Donne is long. We all remember Franklin's "God helps them that help themselves," his "Early to bed and early to rise . . . " and his "Plough deep while sluggards sleep." Ralph Waldo Emerson was able to say it somewhat more elegantly: "A creative economy is the fuel of magnificence." The propriety – indeed, the sanctity – of economic progress was not seriously questioned until our own time. With one notable exception.

John Stuart Mill, who died one hundred years ago, was wildly out of step with his time in one sense at least. Unable to accept that the wealth of nations was governed by immutable universal law, he chose to suggest the heretical notion that sheer material progress might not be the best of all possible goals.

With rare sensitivity and the intellectual courage of a Malthus, Mill suggested that when we cease our preoccupation with "the art of getting on" materially, there might emerge another kind of civilized life style. He questioned the desirability of population growth and economic growth. "I sincerely hope, for the sake of posterity, that they will be content to be stationary before necessity compels them to it. A stationary condition of capital and population implies no stationary state of human improvement. There would be as much scope as ever for all kinds of mental culture and moral social progress, as much room for improving the art of living. . . . " Mill, like Malthus, could not anticipate the technological revolution of the twentieth century, nor the enormity of the population crisis, but his insights were sound.

Stranger still were Mill's grave doubts about the desirability of "every rood of land brought into cultivation which is capable of growing food for human beings, every flowery waste or natural pasture plowed up, all quadrupeds or birds which are not for man's use exterminated as his rivals for food, every hedgerow or superfluous tree rooted out, and scarcely a place left where a wild shrub or flower could grow without being eradicated as a weed in the name of improved agriculture." It was Mill's persuasion that such events need not be inevitable. Things are subject to direction. Decisions could be made, and changes in trends could be implemented. He suggested that population growth could be curtailed. He was strikingly contemporary in his view that once a "stationary" state of society was achieved, there would be the means, through improved methods of distribution of products rather than the sheer quantity of production, for a reasonable degree of comfort for all.

Like Mill, Max Weber brought new dimensions to the examination of economic systems, dimensions which are of critical importance to today's environmentalist. Both were strongly influenced by social factors, and Weber was at pains to emphasize the plurality of causes in the evolution of economic tradition. Like the contemporary ecologist, he recognized the effects of a multiplicity of variable forces playing upon streams of happenings. No event was likely to have one simple origin. In a socio-economic system, as in an ecosystem, many influences are so intimately interrelated that to look for single motive causes is an empty exercise.

Weber did, however, identify a climate of acquisitiveness in Western European culture in which most men were inescapably trapped. He identified too an objective and indifferent (Judaeo-Christian) hierarchy of organization – the bureaucracy – which is a "mechanism founded on discipline," and has little to do with the free and open choice of the individual to determine his own destiny. Locked in the bureaucracy, a man goes on producing as the machinery dictates. This was considered by some, such as Thorstein Veblen, as the normal pattern. He believed that there is in man an innate desire to produce, and that the motive was simply an "instinct of workmanship," not profit. This tendency obviously played into the hands of the industrial system. That there is little of the innate or instinctual in human behavior, Veblen could not know, but he was as modern as today in his observation of the "inordinate productivity of the modern machine process."

Inordinate productivity is what present environmental issues are concerned with – not capitalism or socialism. Environmental issues must not be confused with ideological issues. From the conservationist's point of view, there is no shred of difference between capitalist and socialist societies so long as both stand for inordinate industrial growth and productivity. Industrialization at an increasing rate is the goal of all of the super-powers and their satellites today, and industrialization (including growth in both production and consumption) is the grail of all forms of government in the "developed" world.

An article by Marshall I. Goldman, *The Convergence of Envi-*

ronmental Disruption, makes the point: "By now it is a familiar story – rivers that blaze with fire, smog that suffocates cities, streams that vomit dead fish, oil slicks that blacken seacoasts, prized beaches that vanish in the waves, and lakes that evaporate and die a slow smelly death. What makes it unfamiliar is that this is a description not only of the United States but also of the Soviet Union." Goldman goes on to say that Lake Erie and Los Angeles have no corner on environmental degradation; the "debates and dilemmas are the same" for Lake Baikal and Tbilisi.

After a review of current environmental difficulties in Russia, Goldman concludes that "if the study of environmental disruption in the Soviet Union demonstrates anything, it shows that not private enterprise but industrialization is the primary cause of environmental disruption. This suggests that state ownership of all the productive resources is not a cure-all. The replacement of private greed by public greed is not much of an improvement." In other words, there is little if any basis for the belief that differing economic systems have different degrees of impact on the environment so long as those systems are based on increasing industrialization. It is common nowadays to hear demands from various political persuasions and ideologists couched in ecological terms and directed toward the nationalization of industry. But there is a common bond between them all. All stand for the growth ethic and Calvinistic "progress." Those who are most vocal about the nationalization of Canadian non-renewable resources, for example, are operating from a bias which is equally at odds with sane environmental management as the system they are attacking. When nationalization of primary resources comes, as it inevitably must, it will be on grounds which are ecologically oriented. The grounds will not be ideological.

John Kenneth Galbraith reminds us that in the main (and, one would assume, despite their politics), economists are the natural allies of the industrial system. (*The New Industrial State* should be required reading for all environmental students.)

Galbraith, who carries on the good iconoclastic tradition of J. S. Mill, and who is an equal embarrassment to his colleagues, has convincingly described the identification of the industrial system with the goals of society. If society chooses to limit itself to merely economic goals, says Galbraith, then it is only natural that the industrial system should dominate the state and that the state should in turn serve the ends of the industrial system. That is the bind into which contemporary Western society has got itself, with the industrial system having a virtual "monopoly of social purpose."

That the American style of industrialization today dominates the Western world and shapes the aims and ambitions of much of the non-Western world is one of those accidents of history, the product of a plurality of coincidences. What has happened is not at all unlike the result of any random shift of genetic material, influenced by chance environmental changes.

The plurality of causes, in their combination, became a phenomenon more combustible than any the world has ever seen. As Lewis W. Moncrief has described it, "America is the archetype of what happens when democracy, technology, urbanization, capitalistic mission, and antagonism (or apathy) toward natural environment are blended together." And running through the entire story, like a tight-drawn thread of the finest steel, is the spirit of John Calvin.

CHAPTER FIFTEEN

The Power Structure

As a world species, we now find ourselves at the pinnacle of our ambitions. Man stands at the apex of an ancient and immensely complicated power structure over nature. The structure is man's, not God's; the ambition and the struggle – and the self-appointed mission to achieve it – were man's.

It has been a long and complex journey. The direction was initially the random result of evolutionary whim and chance environmental circumstance. Later, it caught the heady tide of tradition. Throughout, the pilgrimage has been accompanied by grave hardship, not only for men but also for other species and for a multitude of natural environmental types which perforce had to give way before the ultimate mammal. The journey has been an exemplary illustration of the opportunism which is characteristic of evolutionary events. Among the building blocks of the completed power structure there are the fossils of all the species that surrendered their living substance before us, and there are the shells of all the habitat types, large and small, that were eliminated or changed beyond recognition in the process. All this took place in a fraction of an instant of evolutionary time.

The journey of man has not been "purposeful." There were no built-in goals, any more than there are built-in directions for the evolution of a new kind of green heron or song sparrow. Random genetic and environmental events brought us to a stage at which our minds, and thus our cultures, took over. That, in point of time, was yesterday. The day *before* yesterday, changes were as always accidental and at random. The blinding speed at which cultural evolution proceeds tempts us to forget the slow process which is biological evolution. Cultural evolution represents the last twist on a curling metal "slinky-toy" of infinite length. But since we cannot see – or remember – beyond that final cultural spiral in the series, we assume it was ever thus, and anything that happened in prehistoric times has no relevance to our behavior.

Piece by piece, bit by bit, long before we could even formulate the notion, the power structure over nature grew and strengthened. The final "clincher" came with modern technology, which made possible the subjugation of America in two hundred years. The power and the mastery of Bacon and Descartes had become progress in the economic sense, and all had been sanctified by the Protestant church, most especially in its New World, Calvinist form.

As our minds are limited in the number of dimensions we can contemplate, it is useful to picture the power structure as a great latticework in a shape like that of the Eiffel Tower. Man stands at the top. Below him are all the nonhuman components of the biosphere, in descending order according to their relative value to man, which is the sole measure of their cosmic significance. Thus, man stands on Western technology, and Western technology stands on energy resources. Below energy resources there are non-renewable resources. The next stratum downward consists of those resources such as soils, forests, fish and wildlife which we used to call "renewable." We are now a very long way down on the hierarchical scale. Of less importance than renewable resources, and thus of less validity, are "non-economic" animal species. Below the animals which we consider valueless are those plants which have no economic "purpose." Further down is terrestrial wilderness. Of least importance is marine wilderness.

The very word "resource" trumpets Old Testament self-centeredness – a thing nonhuman but useful to man. If it is not useful to man, it is not a resource. It is not, in fact, anything. Resources, on the other hand, have been specially provided as our manageable "harvest."

When the "harvest" principle first evidenced itself in such practices as wildlife management, it was a positive and beneficial concept. It emphasized that one should not, in taking wildlife, eat into capital – or at least not enough to endanger next year's production. The harvest principle is still used, however, in a time when we have learned that there is no harvestable surplus of anything, and there never was. A harvest is a planted, domesticated crop – a stand of wheat or field of tomatoes. For one species of living thing to regard another as a "crop" seems to be the very height of egocentricity and the depth of insensitivity. Before man was in North America to "harvest" the ducks and the balsam fir, the natural system did an exemplary job of feeding these things back into its cycles. If – and it is a big "if " – we can conceive of "harvestable surpluses" in nature, these went to sustain the other component species in the natural community. "Excess" ducks fed pike and prairie falcons, and excess balsam fir fed moose in the winter. Since there was no harvester, there was no harvest, and the wildlife community functioned sublimely well in their absence.

Harvest, resources, management – these concepts have no existence outside our self-centered minds. None is real, except when they are used as justification for further prosecution of the holy war on nature. Like the power structure itself, they are sustained by assumptions which are the result of our most ancient traditions.

Since it is clear that second only to man himself on the Great Scale of Perfection is his Western technology, we are obliged to look at a phenomenon which has been carefully explored by a long list of students in the present decade. It seems that a generalization one can make about technology today is its mindlessness. In the present day, technology proceeds for its *own* sake, not even for the human sake, in a succession of self-fulfilling prophecies. Technology makes something possible, then technology *must* go ahead and do it, regardless of necessity or consequence. As Buckminster Fuller says, "The horrifying truth is that, so far as much technology is concerned, no one is in charge."

Technology in its present form is something quite new. What was originally a kit of increasingly useful tools for helping us get things done has begun to proliferate for the sake of proliferation. Fuller reminds us that in both its capitalist and communist form, industrialization was directed to the advancement of material welfare: "Thus, for the technocrat, in Detroit as well as Kiev, economic advance is the primary aim; technology the primary tool. The fact that in one case the advance redounds to private advantage and in the other, theoretically, to the public good, does not alter the core assumptions common to both. Technocratic planning is *econocentric.*" Technology is the inevitable and indispensible handmaiden of economic growth through industrial production. In today's world – if it is true that we are entering a post-industrial phase – technology having achieved its own momentum, proceeds in its own directions, irrespective of human requirements.

Many conservationists have come down hard on the "technocrats." It may be, however, that we blast the engineers and their accomplices somewhat too freely. One is not quite so Christlike as to intercede for their forgiveness, but it is obvious that they know not what they do. Who has been at pains to educate or even to encourage them otherwise? Who has troubled to inject even the slightest germ of comprehension of man *as* nature into schools of forestry, engineering, or architecture? Who has attempted to suggest to such institutions that there are ecological ground rules, much less ecological ethics? If technical men have so firmly welded and riveted together the supranatural tower of human dominance, then like all technical men they did so at someone else's behest. The fault has never been in our stars. It has been in our traditions.

The majority of our schools are still immured in traditions which are no longer merely irrelevant, but destructive. As Paul Ehrlich has said, "Our educational system has seriously straightjacketed the economic thinking of most of our citizens to the point where they are mesmerized by the axiomatic 'good' of growth." We have encouraged – even directed – students to look to growth for the ultimate satisfaction and to look to technology for the achievement of growth. The ecological innocence and indifference of a high proportion of graduates from most traditional schools is appalling. Our educational system is no breeder of environmental humility.

The power structure has become critical to the entire biosphere in our time because of the exponential increase in our *ability* to achieve the conquest of our environment. The goal has always been there; only the methods have changed. Modern tools have allowed us to accomplish what we would certainly have accomplished in earlier times, had we had the techniques.

It is true that our contemporary picture of "progress" is relatively recent; in its present form it dates back no further than the seventeenth century. But the blueprint for the tower of Babel existed long before that. Man and nature have been separate in men's minds since long before there was any notion of industrial progress. They were conceptually separate in a time when all things were seen as static and immutable. There was a pyramid even then. Teilhard's tree is at least as old as Aristotle, and power over nature goes back to faintly magical stirrings in a deep and lightless Neanderthal cave.

Although Charles Reich's description of today's elite (in *The Greening of America*) is more of an indictment than a portrait, it is impossible to disagree with him. The elite, he says, have been cruelly deceived into believing that "richness, satisfactions, joy of life, are found in power, success, status, acceptance, popularity, achievements, rewards, excellence, and the rational, competent mind." They want "nothing to do with dread, awe, wonder, mystery, accidents, failure, helplessness, magic." Such people, who make up the contemporary "establishment," Reich says, have produced a society that is "the image of its own alienation and impoverishment." He goes on to remark that there is an "imperative logic" behind the destruction of nonhuman nature as practiced by the Corporate State, because the state "draws its vitality by a procedure that impoverishes the natural world." We have always had that imperative logic; it has been part of the evolving power structure since prehistory. Today, our technical efficiency has matched our aspirations. The danger is that the technological machine, in its mindless acceleration, may now run away from *all* of nature – which includes ourselves.

I have entertained a vision of a sea cucumber – a marine invertebrate belonging to the animal phylum called echinoderms, the spiny-skinned ones. Echinoderms are usually characterized by five-part symmetry, as with the legs of a starfish or the segments of a sea urchin. Like the others in its phylum, and unlike all other animals in the world, the sea cucumber's vascular system is filled not with blood, but with water. It is a sluggish and slow-moving animal which lies more or less immersed in muck at the bottom of the ocean. It takes sludgy water in through its mouth, extracts whatever nutrients the water contains, and voids water and unusable sand at the other end. The sea cucumber, as it blindly pulses on the ocean floor, does not select its food; it ingests whatever happens to be in front of it in the watery sediment, and it voids whatever it cannot use.

There is a striking analogy between the sea cucumber and the insensate ingesting, processing and voiding of modern industry. Like the echinoderm, industry does not care a hoot what it takes in at one end, and cares even less about what it deposits at the other, so long as it can extract something of value in the process. Squeeze a sea cucumber and you get water, not blood. Squeeze industry and you do not get blood either. Squeeze the echinoderm a little harder, and the animal will undergo a violent paroxysm in the course of which, according to Rachel Carson, it will "hurl the greater part of its internal organs through a rupture in the body wall. Sometimes this action is suicidal, but often the creature continues to live and grows a new set of organs." Squeeze industry hard enough and it, too, will behave as though you are killing it.

Since neither sea cucumber nor industry appears to take the slightest interest in what flows in the front door and flows out the back, so long as it is nourishing in the process, one can take a moment of frivolity to indulge in the alchemist's fantasy. What would the organism need to ingest in order to cause it to excrete pure gold? Or perhaps pure air. All that would be needed is the right formula. It must be said of the sea cucumber that it voids almost nothing but the finest and cleanest sand. Moving sand from place to place is an innocuous activity which may even be beneficial in the long run. The benefits of most industrial processing today, apart from those involved with basic necessities, is another question.

However, like the sea cucumber, the industrialist is not exercising willful choice. He does the only thing he knows how to do, and he does it because he has always done it. He differs from the echinoderm only in that he believes what he does to be right and proper. Both "do their thing" because that is what they have evolved to do within the context of their special environments. Both would have disappeared onto the slag-heap of misfits long since had their environments not permitted them to perpetuate themselves.

No man is so far removed from nature as the liberally educated humanist, because the cosmos centers on his mind, and the mind of man is the measure – and the envelope – of all things. This is the man to whom, as Paul Shepard says, nature is either natural resources or scenery – or else it is that endlessly fascinating garden of delights called "human nature." The run-of-the-mill humanist is incredibly ignorant of, and thus indifferent to, his biological context; and he is even somewhat reluctant to be reminded of it. The liberal humanist is dangerous to the biosphere, and thus to mankind, because by and large he is the leader, the educator, the opinion-maker. Even in a money-oriented society, his influence exceeds that of the technocrat or the industrialist. His "culture" grants him access to seats of power and influence. He is the key to the entire supranatural pyramid, because he is ancient anthropocentricity in its most highly developed form.

The humanist is the very picture of tolerance and compassion toward nature. He supports "humane" societies, he takes his children to the zoo to see the amusing antics of captive "lower" animals, and to the museum to see curiosities from the farthest corners of man's world. He is interested in everything, because he has a cultivated mind. He is gentle, genteel, and humanitarian. Unlike men in trade, commerce and the technical professions, the liberal humanist has been at an evolutionary standstill for over two thousand years.

That is a regrettably long time, in view of the potential speed of cultural evolution by comparison with the more deliberate pace of biological processes. While most other aspects of our culture were undergoing remarkable transformations and developments, the traditional humanities remained locked in antiquity – whether in a Romantic fixation on things past or in a protective shell is open to opinion.

Although one is chagrined at the continuing failure of the humanities to enter the contemporary arena of environmental issues, and at their apparent insistence on maintaining the rigid integrity of the conceptual separation of man and nature, it is nonetheless in the humanities that we must concern ourselves.

If "human ecology" is ever to emerge as a definable body of knowledge or area of investigation, it is far more likely to emerge from the humanities than from any of the hard or soft sciences. Man is the cultural animal. Culture created the power structure over nature, and only in culture is the blueprint for its dismantlement.

CHAPTER SIXTEEN

The Graceful Continuity

Except for rare aberrations, the humanistic tradition has never been reconcilable with the realities of living on Earth. There is a critical need for a dispassionate and thoughtful review of the basic assumptions that underlie our literature, philosophy, and arts. Only in this way can we begin to pry open the shackles of closed thinking that have plagued Western culture for so many centuries. This is where artificial selection comes in. We develop fancy pigeons by selecting for breeding those birds with the attributes we desire. The others we suppress. The same can be done with component parts of our culture. But before we can select those components for propagation or rejection, we must be able to identify and evaluate them within the complex "ecosystem" which is our collective accumulated thought.

Values have been variously defined, but a common element appears to be the perception of some degree of merit, or at least of usefulness, in the eye of the possessor of the value in question. Values are also defined as qualities, standards, or even principles, which are considered by the possessor as desirable, or at least worthwhile. These perceptions and evaluations are of course no more than indications of what goes on inside the head of the individual beholder, or valuer. They have no intrinsic "value," no intrinsic reality. The reality is in what we *do* on the basis of our values. In a sense, a value is at the same time a *motivation* to achieve the satisfaction and pleasure one "values." Thus a value becomes a *reason*. Like energy and matter, satisfaction and motive become at once the same and inseparable.

Institutions are, among other things, reflections or manifestations of sets of currently prevailing values. An obvious example is the GNP. The GNP is an institution; even today, relatively few people question its value as a measure of progress. Progress itself is an institution. The conceptual separation of man and nature, and the conceptual power structure over nature, are institutions which reflect prevailing values. The power structure itself is built on other institutions – religious, philosophical, economic, political, humanistic. These are traditional "givens"; they are reflections of subjective value systems making up the largely artificial "construct" which is contemporary Western culture.

Institutions such as the self-perpetuating profit system and the gospel of individual freedom are additional evidence of the value systems we have inherited. They are firmly ingrained in us; to change them would be difficult. The acceptance of any new set or sets of values would involve a change in "mental set" of a magnitude and profundity which would only be comparable to the intellectual and emotional "conversion" implicit in certain kinds of religious experience. Certainly it would have to be on a grand scale. But it could happen: there are precedents.

Even very radical shifts in value systems are possible if there are good reasons for them. The fact that values are often their own reasons can help to make the transition smoother. One thinks of times of national emergency, such as war, when the value of individual freedom becomes relatively unimportant in the face of the greater social purpose. Values such as the sanctity of human life evaporate on the instant. Even the value of profit-making (to at least some extent) and that of fair competition are secondary to the war production effort. The reason is the prosecution of the war, and the prosecution of the war becomes the new value. The institution of democracy is temporarily, and quite voluntarily, suspended.

There can also be shifts in value systems when there are reasons such as the relative abundance or scarcity of commodities or even objects of interest. Society is quite prepared to value the whooping cranes or the blue whales when it is informed that there are so very few of them. It is civilized to value things that are in short supply. If they are few, they must be valuable. Society is somewhat less inclined to value things such as raccoons or redwinged blackbirds because there are still lots of them.

There have been changes in institutionalized values and value systems when developments in popular thought have declared them to be irrelevant or archaic. In our own time we are seeing at least some changes in the values formerly attached to such institutions as wedlock, motherhood, militarism, conspicuous consumption, the old-time religion, haircuts, bras and neckties. If people feel better without these accoutrements, then their absence is the new value. In their place have been substituted such things as chronic skepticism, blue jeans, compulsive sharing, and whatnot. Whether the values of "Consciousness III" will persist is of course unknown. What *is* known is that spiritual bankruptcy is in institutions, not in people.

In their natural environments, living beings face an infinity of survival problems – food shortages, predators, diseases, competitors, population stresses, and so on. The gravest problem they now face, however – man's self-appointed supremacy over them – is itself strangely like an ecosystem. It has a vast and complicated array of interlocking components. All components interact with each other. All pull and push and pulse with rhythmic tensions, as in a dynamic web of nature. The conceptual power structure is so old, its foundations so embedded in time and tradition, that it has become a very real community of mutual self-interest, sustained and reinforced by its interacting components. Its homeostasis is maintained by the constant energy exchanges and mutual feedbacks between myth and superstition, religion and philosophy, science and technology, politics and economics, fear and arrogance. All, for their symbiotic survival, have depended upon all the others in the creation of a complicated web of interdependent strands which has now become a conceptual monolith of terrifying scope – and stability.

But the very fact of the structure's stability means it is not static. Like any other ecosystem, it is in constant flux. Also, as in the web of any other ecosystem, pressure on one strand is transmitted in greater or lesser measure to all the others. Pull on the strand of traditional, classical economics, and you get

immediate political oscillations. Gently nudge the strand of
education, and you get sympathetic responses from unexpect-
edly distant branches of religion. Pluck the thread of progress, and violent are the paroxysms throughout.

As any naturalist knows, the quickest and neatest way to
destroy an ecosystem is to simplify it, to reduce its complexity
and thus to short-circuit the equilibrium maintained by the
mutual interdependence of its component parts. Perhaps the
traditional, cultural, institutional, conceptual eco-construct
can be decomplexified by our deliberate manipulation – by
the exercise of our conscious choice. Intervention in its work-
ings will require degrees of courage, sacrifice, imagination
and generosity which have not frequently been displayed in
the course of man's relationship with his environment. One
hesitates to predict whether we will be willing to undertake it.
The destruction of the power hierarchy over nature will require a shift in attitudes more profound than we can presently
imagine.

The process of simplification or decomplexification will
be drastic. Suppose one were to elect to have an initial go at
the "rights of man" – the God-given rights of man the individual and man the species. Suppose it were feasible to
actually remove some of those rights, one after the other. The
consequences might be astonishing. Let us think of such traditions as every man's right to a plot of land with a house on
it, the land to do with as he will. A man can do as he wishes
with his own property – his trees, his water, his soil, and so on.
We now know, however, that a man can no longer pollute as
he pleases. He is controlled in that. He will soon be controlled
much more stringently in his use of the soil itself. A man
should no more be allowed to own the living soil than he now
owns the air he breathes. Environmental forces are already
eroding traditional "rights."

Then there is the right to have children. Suppose people
were no longer permitted to reproduce beyond the replacement level. Replacement means one adult, one child – zero
population growth. In more practical terms, it means two

children per female human being. Is the imagined right to have as many children as one pleases really in the best interest of all mankind? One might go further, and say that there are some people who do not have the right to have children at all. There are other people who are quite capable of producing intelligent, strong, beautiful children, but whose right to *raise* them might be gravely questioned. I know children of immense promise who are being destroyed by parents of eighteenth-century persuasions. The environmentalist must look hard at traditional human freedoms.

There are other "rights" such as the imagined right of men to kill nonhuman animals for amusement. Clearly the environment itself will deal with this tradition, simply as the effect of men having joyfully massacred so many ducks, geese, rhinos, elephants and Cape buffalos that there will not be enough of them to go round. A similar end will come to the fashion industry's apparent determination to exploit to the bitter end the final stocks of leopard, tiger, jaguar, and the rest.

What of the more fundamental, unquestioned rights of man the species? The right to populate at will must certainly be removed, either by our conscious choice or by a natural backlash on the part of the biosphere itself. The right to dominate animals of other species, and to dominate landscapes, will not be subverted as readily. Other beings, as species and as landscapes, do not have the "clout" of the combined forces of the biosphere. But that right, too, will disappear. It will be a sad process, for we will not give up the right to dominate without a struggle – a struggle which will cost both human and nonhuman nature exorbitantly.

It will not be in our best interests to allow the environment to dismantle our conceptual power structure for us. In such an eventuality, cosmic forces would make life devastating. We should not expect the environmental counterattack to be nearly so dramatic or spectacular as the ancient vision of the Apocalypse, but it would be equally disastrous. Because it would not be sudden, it would be even more agonizing.

It has never been difficult for us to see the necessity for and to devote abundant energy to the inflicting of our ideologies on people who because of their circumstances are not privy to our insights. It should not be difficult for those of us who have been privileged to receive "the word" about the deteriorating *human* condition to inflict that word, by one means or another, on those who have not got it. We have the evidence. Unlike redemption, or any political "ism," the environmental crusade has plenty of hard fact to back it up. There is no mystery about it. The threat of hellfire is no longer veiled; there are millions of starving people and there are billions of mutilated nonhuman beings to bear witness to the dislocation of the biosphere.

One of the saddest ironies is human adaptability. As a mammal, man has an internal heat-regulating system and can live in a variety of climates. As a primate, man is a sort of generalist and can live in a wide variety of habitat types. As a technologist, man can live virtually anywhere. More than that, however, is man's psychic adaptability. We can persuade ourselves of almost anything, however fanciful – such as our separation from nature. Worse, we can get used to living under the most trying conditions. People suffered abominably in Hitler's concentration camps, but a good many made a sufficient psychic adjustment. People live today in indescribable filth and squalor, without enough food, without medical assistance, without hope. But their psychic adaptability allows them to *live,* and millions of them live long enough to beget millions more. People in large industrial cities get used to levels of air pollution which choke and nauseate new arrivals from the "wide open spaces." We get used to these things, just as we get used to the discomforts of aging, because they are gradual developments, and we are rarely aware of what is happening until it has happened. Adaptability can be insidious.

Adaptability can produce startling results for a species – and in the psychic sense, even for an individual – but one thing that adaptability can never do is guarantee immortality.

Immortality has yet to be demonstrated in the cosmos, and it has yet to be demonstrated on Earth. Certainly it will not be demonstrated in the lifetime of the species man. No cell, nor individual, nor species, is immortal. There is comfort in that fact. Eternal life, like life imprisonment, would be the cruellest punishment that could be inflicted on a living being.

One wonders why the dream of immortality has always been so important to us as individuals, as societies, as cultures, and as a species. The naturalist is not all that "hung up" on immortality. I made my peace with that issue years ago (and it *is* a peaceful thought). Saul Alinsky was recently quoted in *Playboy*: "Once you accept your own mortality on the deepest level, your life can take on a whole new meaning."

Earth will continue in its orbit for the balance of its own lifetime regardless of present and future human activities. Life processes and their manifestations may be simplified by contemporary standards, with organisms of a somewhat lesser degree of overt complexity of organization filling the initial void which our last convulsions will have left. But the overall background complexity – the programming – will remain. Some of Earth's matter may by that time have been released as energy by nuclear devices, but only a miniscule amount. Most of the matter that is here today will still be here. The essential atoms may be in somewhat different molecular combinations, but since nothing goes away, all will be here. In whatever form we leave them, the basic elements will eventually sort themselves out into new, workable structures which will be adapted to the new, posthuman biosphere.

Man *might* take with him more than he should of the necessities for posthuman life on Earth. The ultimate blasphemy would be man's total destruction of the materials and possibilities of the life process. This, I reassure myself, is impossible. Man will be prevented from committing this final and ugliest of acts by the sheer resilience of the biosphere. It is beyond our capacity to destroy everything. In fact, it is beyond our capacity to destroy *anything* – merely to rearrange its parts.

The appeal of mortality is in the graceful continuity which

it makes possible. Can you picture anything more dull and depressing than an inert, static, unchanging world, with all things just as they are today? The glory and the grandeur are in the flux, the constant shifting which is made possible by the combination, decomposition and recombination of the basic building materials of life. Though I do not expect that I shall be reborn directly as a crocus, I know that one day my atoms will inhabit a bacterium here, a diatom there, a nematode or a flagellate – even a crayfish or a sea cucumber. I will be here, in myriad forms, for as long as there are forms of life on Earth. I have always been here, and with a certain effort of will, *I can sometimes remember.*

We attempt to distinguish between past, present, and future when no such distinction is possible. Those spinning particles that at this instant are John Livingston have always been around, somewhere, and always will be. At this instant, quite by chance, they happen to be in their present configuration. We attempt to distinguish between plants and animals, but the harder we press our determination to differentiate between them, the fuzzier the distinction becomes. We attempt to distinguish between living and nonliving, and then we stumble on the virus. We attempt to distinguish between man and other living beings, when there is no distinction. Differentiation is the cruelest block to comprehension of the oneness of being.

Oneness and the notion of timelessness are hard on mental images of purposes and goals, on pictures of causes and effects, on visions of destinies. The ancient Greeks knew better about these things than did Francis Bacon, and they knew much better than the contemporary decision-maker raised in the Western humanistic tradition. Not only did the Greeks know about oneness, they knew also about *hubris.* Alexander Pope knew about the error of reasoning pride, and Shakespeare about the relative significance of the poor posturing player.

Time is fast running out for the dismantling of the institutions which have kept us so grimly locked in step with "progress." There is even less time for reflection on the merits of the traditional components of our culture which have brought us – and all of nature – to the present point of departure. A point of departure it is, either from the narrow and egocentric cultural course we have adopted, or premature departure from the blue planet itself. If we are not yet capable of identifying the specific threads in the fabric of our beliefs which have sustained the entire tapestry upon which the myth of human dominance is emblazoned, then it may be too late already.

The hope for survival of nonhuman nature is dim. There is a familiar scenario. As conditions worsen for human populations – as they will, initially, in underprivileged parts of the world – every ounce and erg of our most refined technological skills and energies will be brought into play to extract from Earth and its nonhuman inhabitants the basic ingredients for human survival. We will first destroy all of the larger animals, either for meat or because they compete with us for space, together with those which may be intolerant of our activities because of their specific natural specializations. Extinction of nonhuman species, without replacement, will continue at an accelerating rate, until the only nonhuman living beings remaining will be those who are willing to share their squalor with us – rats, gutter curs, and parasites and micro-organisms which thrive in times of environmental dislocation.

Our capacity for seeing into the future is limited – and we do not want to know about futures of that kind. We withdraw behind the opaqueness of closed imaginations and familiar fancies. We acknowledge that, yes, the situation is bad, but human ingenuity, creativity, enterprise and goodwill can overcome all difficulties.

While we should be unraveling the threads of tradition, we are weaving ever more elaborate curtains of rationalization. Every avenue of questioning closed off is another route to intellectual and spiritual freedom barricaded forever.

There is no engineering answer to a problem created by culture. The worst in humanistic ways of thinking opened and kept open the conceptual man/nature dichotomy, and only the mature wisdom and insight that characterize the best in the natural philosophic tradition can mend it.

BIBLIOGRAPHY

Adamson, George, *Bwana Game*. London: Collins & Harvill Press, 1968.
Ardrey, Robert, *African Genesis*. New York: Atheneum, 1961.
—— *The Territorial Imperative*. New York: Atheneum, 1966.
—— *The Social Contract*. New York: Atheneum, 1970.
Babbitt, Irving, *Rousseau and Romanticism*. Cleveland, Ohio: Meridian Books, 1955.
Bates, Marston, *The Forest and the Sea*. New York: Random House, 1960.
Bernal, J. D., *Science in History* (4 vols.). Harmondsworth, Middlesex: Pelican Books, 1969.
Berrill, N. J., *Man's Emerging Mind*. New York: Dodd, Mead & Company, 1955.
—— *You and the Universe*. New York: Dodd, Mead & Company, 1958.
Bertram, Colin, "Man Pressure." *Oryx*, 7 (2-3): 97-101, August, 1963.
Biggar, H. P. (ed.), *The Works of Samuel de Champlain*. Toronto: The Champlain Society, 1936.
Brower, David (ed.), *Wilderness: America's Living Heritage*. San Francisco: Sierra Club, 1961.
Buchan, John (ed.), *A History of English Literature*. London: Thomas Nelson and Sons, Ltd., 1923.
Buettner-Janusch, John (ed.), *Evolutionary and Genetic Biology of Primates* (2 vols.). New York: Academic Press, 1963.
Burrill, Donald R. (ed.), *The Cosmological Arguments*. Garden City, New York: Doubleday (Anchor Books), 1967.
Carson, Rachel, *The Edge of the Sea*. Boston: Houghton Mifflin Co., 1955.
Chard, Chester S., *Man in Prehistory*. New York: McGraw-Hill, 1969.
Clark, Kenneth, *Civilisation*. London: British Broadcasting Corporation and John Murray, 1969.
Clement, Roland C., "The Ethics of Survival." *The Ohio Journal of Science*, 69 (1): 15, January, 1969.
Coale, Ansley J., "Man and His Environment." *Science*, 170: 132-136, 1970.
Commoner, Barry, *The Ecological Facts of Life*. U. S. National Commission for UNESCO, 1969.
—— *Science and Survival*. New York: Viking Press, 1967.
Dansereau, Pierre, *The Hope of Human Ecology*. Canadian Commission for UNESCO, July, 1969.
Darling, Frank Fraser, *Wilderness and Plenty*. New York: Ballantine Books, 1970.
Deevey, Edward S., Jr., "The Human Population." *Scientific American*, 203 (3): 195-204, September, 1960.
Disch, Robert (ed.), *The Ecological Conscience*. Englewood Cliffs, New Jersey: Prentice-Hall, Inc., 1970.
Dorst, Jean, *Before Nature Dies*. Boston: Houghton Mifflin Co., 1970.
Dubos, René, *The Torch of Life*. New York: Simon and Schuster, Inc., 1962.
—— *So Human an Animal*. New York: Charles Scribner's Sons, 1968.
—— *Reason Awake*. New York: Columbia University Press, 1970.
Ecology Action East, "The Power to Destroy, the Power to Create." *Rat*, January, 1970.
Edberg, Rolf, *On the Shred of a Cloud*. University, Alabama: University of Alabama Press, 1969.
Ehrlich, Paul R. and Anne H., *Population Resources Environment*. San Francisco: W. H. Freeman and Company, 1970.
Eiseley, Loren, *The Immense Journey*. New York: Random House, 1957.
Elder, Frederick, *Crisis in Eden*. Nashville, Tennessee: Abingdon Press, 1970.
Elliott, J. H., "Spanish Holocaust." *The New York Review of Books*, November 5, 1970.
Falk, Richard A., *This Endangered Planet*. New York: Random House, 1971.
Firsoff, V. A., *Life, Mind and Galaxies*. London: Oliver and Boyd, Ltd., 1967.

Forstner, Lorne J., and Todd, John H. (eds.), *The Everlasting Universe*. Lexington, Mass.: D. C. Heath and Company, 1971.

Frazer, J. G., *The Golden Bough*. London: Macmillan, 1922.

Fuller, R. Buckminster, *Utopia or Oblivion: the prospects for humanity*. New York: Bantam Books, 1969.

———— *Technology and the Human Environment*. U. S. Senate Subcommittee statement, March 4, 1969.

Galbraith, John Kenneth, *The New Industrial State*. Boston: Houghton Mifflin Co., 1967.

Goldman, Marshall I., "The Convergence of Environmental Disruption." *Science*, 170: 37-42, 1970.

Goodman, Paul, "Can Technology be Humane?" *The New York Review of Books*, November 22, 1969.

Grant, George, *Technology and Empire*. Toronto: House of Anansi, 1969.

Hardin, Garrett, *Nature and Man's Fate*. New York: Holt, Rinehart and Winston, Inc., 1959.

Heilbroner, Robert L., "Ecological Armageddon." *The New York Review of Books*, April 23, 1970.

Hewes, Gordon W., "Food Transport and the Origin of Hominid Bipedalism." *American Anthropologist*, 63: 687-710, 1961.

Hockett, Charles F., and Ascher, Robert, "The Human Revolution." *Current Anthropology*, 5(3): 135-168, 1964.

Hopkins, David M. (ed.), *The Bering Land Bridge*. Stanford, California: Stanford University Press, 1967.

Howell, F. Clark, and the Editors of Time-Life Books, *Early Man*. New York: Time, Inc., 1965 (revised 1968).

Howells, William W., "Homo Erectus." *Scientific American*, 215: 46-53, November, 1966.

Hoyle, Fred, *The Nature of the Universe*. Harmondsworth, Middlesex: Pelican Books, 1963.

Huxley, Julian, *Essays of a Biologist*. London: Chatto and Windus, 1923.

Jackson, Wes, *Man and the Environment*. Dubuque, Iowa: William C. Brown Company, 1971.

Jastrow, Robert, *Red Giants and White Dwarfs*. New York: Harper and Row, 1967.

Josselyn, John, *New-Englands Rarities*. London: G. Widdowes, 1672.

Kruuk, Hans, "Clan-system and feeding habits of spotted hyenas *(Crocuta crocuta)*." *Nature*, 209 (5029): 1257-1258.

Kummer, Hans, *Primate Societies*. Chicago: Aldine-Atherton, 1971.

Laing, R. D., *The Politics of Experience*. Harmondsworth: Penguin Books, 1967.

Leakey, R. E. F., "Fauna and Artefacts from a New Plio-Pleistocene Locality near Lake Rudolf in Kenya." *Nature*, 226, April 18, 1970.

Leopold, Aldo, "The Conservation Ethic." *Journal of Forestry*, October, 1933.

———— *Sand County Almanac*. New York: Oxford University Press, 1949.

Livingston, John A., *Canada*. Toronto: Natural Science of Canada Ltd., 1970.

———— "Man and His World: A Dissent," in *Wilderness Canada*. Toronto: Clarke, Irwin, 1970.

———— "Neo-Nationalism and Ecological Independence – A Personal View." *Ontario Naturalist*, 8(4), December, 1970.

Lovejoy, Arthur O., *The Great Chain of Being: a study in the history of an idea*. Cambridge, Mass.: Harvard University Press, 1936.

———— *Essays in the History of Ideas*. Baltimore, Maryland: The Johns Hopkins Press, 1948.

MacNeill, J. W., *Environmental Management*. Ottawa: Information Canada, 1971.

Margolis, John, "Land of Ecology." *Esquire*, March, 1970.

Martin, P. S., and Wright, H. E., Jr. (eds.), *Pleistocene Extinctions: the search for a cause*. New Haven, Connecticut, and London: Yale University Press, 1967.

Marx, Leo, *The Machine in the Garden*. New York: Oxford University Press, 1964.

Matthiessen, Peter, *Wildlife in America*. New York: The Viking Press, 1959.

232

Mech, David L., *The Wolf*. Garden City, New York: The Natural History Press, 1970.

Merton, Thomas, "The Wild Places." *The Center Magazine*, July, 1968.

Moncrief, Lewis W., "The Cultural Basis for our Environmental Crisis." *Science*, 170: 508-512, 1970.

Montagu, M. F. Ashley (ed.), *Man and Aggression*. New York: Oxford University Press, 1968.

Morris, Desmond, *The Naked Ape*. London: Jonathan Cape, 1967.

—— *The Human Zoo*. London: Jonathan Cape, 1969.

Mumford, Lewis, "Closing Statement." *Future Environments of North America*, F. Fraser Darling and John P. Milton (eds.), The Conservation Foundation, 1966.

McHarg, Ian L., "Values, Process, and Form." *The Fitness of Man's Environment*, Washington, D. C.: The Smithsonian Institution Press, 1968.

McKinley, Daniel, "The New Mythology of 'Man in Nature.'" *Perspectives in Biology and Medicine*, 7(1): 93-105, Autumn, 1964.

Nash, Roderick, *Wilderness and the American Mind*. New Haven, Connecticut: Yale University Press, 1967.

Neill, Wilfred T., *The Last of the Ruling Reptiles*. New York: Columbia University Press, 1971.

Odum, E. P. and H. T., *Fundamentals of Ecology*. Philadelphia: W. B. Saunders Co., 1959.

Odum, Eugene P., *Ecology*. New York: Holt, Rinehart and Winston, Inc., 1963.

Odum, Howard T., *Environment, Power, and Society*. New York: John Wiley and Sons, Inc., 1971.

Ohlin, Goran, *Historical Outline of World Population Growth*. Background paper for United Nations World Population Conference, Belgrade, 1965.

Olsen, Jack, *Slaughter the Animals, Poison the Earth*. New York: Simon and Schuster, 1971.

Orenstein, Ronald I., "Tool-use by the New Caledonian Crow (*Corvus moneduloides*)." *Auk*, 89(3): 674-676, July, 1972.

Paddock, William and Paul, *Famine 1975!* Boston: Little, Brown and Company, 1967.

Potter, Frank M., Jr., "Everyone Wants to Save the Environment." *The Center Magazine*, March, 1970.

Prescott, W. H., *The Conquest of Mexico*. London: J. M. Dent & Sons, 1901.

—— *History of the Conquest of Peru*. London: J. M. Dent & Sons, 1908.

Prince Philip, Duke of Edinburgh, and Fisher, James, *Wildlife Crisis*. New York: Cowles Book Company, Inc., 1970.

Randall, John Herman, Jr., *Aristotle*. New York: Columbia University Press, 1960.

Reich, Charles A., *The Greening of America*. New York: Random House, 1970.

Reinow, Robert and Leona Train, *Moment in the Sun*. New York: Ballantine Books, 1967.

Romer, Alfred S., *Man and the Vertebrates* (2 vols.). Harmondsworth, Middlesex: Penguin Books, 1933 (revised 1954).

Schaeffer, Francis A., *Pollution and the Death of Man: the Christian view of ecology*. Wheaton, Illinois: Tyndale House Publishers, 1970.

Schaller, George B., *The Year of the Gorilla*. Chicago: The University of Chicago Press, 1964.

—— *The Deer and the Tiger*. Chicago: The University of Chicago Press, 1967.

Schaller, George B., and Lowther, Gordon R., "The Relevance of Carnivore Behavior to the Study of Early Hominids." *Southwestern Journal of Anthropology*, 25(4), Winter, 1969.

Schweitzer, Albert, *Out of My Life and Thought*. New York: Henry Holt & Co., Inc., 1949.

Sears, Paul B., "The Inexorable Problem of Space." *Science*, 129: 9-16, 1958.

Shepard, Paul, and McKinley, Daniel (eds.), *The Subversive Science*. Boston: Houghton Mifflin Co., 1969.

Slobodkin, Lawrence B., "Aspects of the Future of Ecology." *BioScience*, 13(1): 16-23.

Snyder, Gary, "Poetry and the Primitive." *Earth Household*, 1967.

Soule, George, *Ideas of the Great Economists*. Toronto: Mentor Books, 1952.

Spencer, Roscoe, "Individual and Species: Biological Survival." *The Humanist,* 3: 155-161, 1958.

Stalker, A. Mac S., "Geology and Age of the Early Man Site at Taber, Alberta." *American Antiquity,* 34(4): 425-428, October, 1969.

Storer, John H., *Man in the Web of Life.* New York: Signet Books, 1968.

Storr, Anthony, *Human Aggression.* Harmondsworth, Middlesex: Penguin Books, 1968.

Teilhard de Chardin, Pierre, *The Phenomenon of Man.* New York: Harper and Row, 1959.

———— *Man's Place in Nature.* New York: Harper and Row, 1966.

Thomas, William L., Jr. (ed.), *Man's Role in Changing the Face of the Earth* (2 vols.). Chicago: University of Chicago Press, 1956.

Toffler, Alvin, *Future Shock.* New York: Random House, 1970.

Trevelyan, G. M., *English Social History.* London: Longmans, Green, 1942.

van Lawick-Goodall, Hugo and Jane, *Innocent Killers.* London: Collins, 1970.

van Lawick-Goodall, Jane, *In the Shadow of Man.* London: Collins, 1971.

Velikovsky, Immanuel, *Worlds in Collision.* New York: Doubleday, 1950.

Washburn, Sherwood L. (ed.), *Social Life of Early Man.* Chicago: Aldine Publishing Company, 1961.

Weber, Max, *The Protestant Ethic and the Spirit of Capitalism.* New York: Charles Scribner's Sons, 1958.

Weisberg, Barry, "The Politics of Ecology." *Liberation Magazine,* January, 1970.

White, Lynn, Jr., "The Historical Roots of Our Ecologic Crisis." *Science,* 155: 1203-1207, 1967.

Whiteside, Thomas, *Defoliation.* New York: Ballantine Books, 1970.

Willey, Basil, *The Eighteenth-Century Background.* Harmondsworth, Middlesex: Penguin Books, 1962.

Wright, Louis B. (ed.), *Newes from the New-World.* Companie of the Friends of the Huntington Library, 1946.

Wynne-Edwards, V. C., *Animal Dispersion in Relation to Social Behavior.* London: Oliver and Boyd, 1962.

———— "Self-Regulating Systems in Populations of Animals." *Science,* 147: 1543-1548, 1965.

INDEX

236